Communications
in Computer and Information Science 998

Commenced Publication in 2007
Founding and Former Series Editors:
Phoebe Chen, Alfredo Cuzzocrea, Xiaoyong Du, Orhun Kara, Ting Liu,
Krishna M. Sivalingam, Dominik Ślęzak, and Xiaokang Yang

More information about this series at http://www.springer.com/series/7899

Yuri Shokin · Zhassulan Shaimardanov (Eds.)

Computational and Information Technologies in Science, Engineering and Education

9th International Conference, CITech 2018
Ust-Kamenogorsk, Kazakhstan, September 25–28, 2018
Revised Selected Papers

Editors
Yuri Shokin (iD)
RAS
Institute of Computational Technologies
Novosibirsk, Russia

Zhassulan Shaimardanov
D. Serikbayev East Kazakhstan State
Technical University
Ust-Kamenogorsk, Kazakhstan

ISSN 1865-0929   ISSN 1865-0937 (electronic)
Communications in Computer and Information Science
ISBN 978-3-030-12202-7   ISBN 978-3-030-12203-4 (eBook)
https://doi.org/10.1007/978-3-030-12203-4

Library of Congress Control Number: 2019932171

This Springer imprint is published by the registered company Springer Nature Switzerland AG
The registered company address is: Gewerbestrasse 11, 6330 Cham, Switzerland

Preface

The International Scientific and Practical Conference "Computational and Information Technologies in Science, Engineering and Education" (CITech) has a long and rich tradition and has been held regularly since 2002.

Historically, the conference was organized in close cooperation between Russian and Kazakh scientists and the general area of discussion was the most advanced achievements in the field of computational technology. The geographic reach of the conference later expanded and now it is attended by leading scientists from Europe, the USA, Japan, India, and Turkey, among others.

The purpose of the conference is the dissemination of new knowledge and scientific advances among the participants. A special feature of this conference is to involve young scientists in the assessment of their scientific achievements through their interaction with the two countries' leading scientists. Participating in CITech has helped formed a community of new-generation young scientists who are currently conducting important research in the field.

CITech has been held in Almaty (2002, 2004, 2008, 2015), Pavlodar (2006), and Ust-Kamenogorsk (2003, 2013, 2018). An important role in the formation of stable traditions for organizing and conducting CITech is played by the personal friendships of scientists from the Novosibirsk Scientific School, such as Prof. S. Smagulov, N. Danaev, Y. Shokin, V. Monakhov, B. Zhumagulov, and many others. Unfortunately, some of them are no longer among us, but we will always remember their contribution to science and education and keep their unforgettable image in our hearts.

For CITech-2018, the call for participation was issued in January 2018, inviting researchers and developers to submit original recent work to the Program Committee. Presentations at the conference took place in five sessions: "Mathematical Modelling of Technological Processes," "High-Performance Computing," "Control, Processing, and Protection of Information," "Automation and Control of Technological Processes," and "New Information Technologies in Education." In all, 86 papers were received from authors all over the world. The Program Committee consisted of 52 members selected from 16 countries. Furthermore, 33 external expert reviewers from the community were invited to help with specific papers. Finally, the committee selected 24 papers for publication, and for presentation in the research paper sessions. We are grateful to the members of the Program and Organizing Committees, the additional reviewers for their help in preparing this publication, and the Ministry of Education and Sciences of the Republic of Kazakhstan for support in the organization of conference. We hope the papers of CITech 2018 will be interesting for the readers and of value for the scientific community.

February 2019

Yuri Shokin
Zhassulan Shaimardanov
Bakytzhan Zhumagulov

Organization

Program Committee

Program Committee Co-chairs

Yuri Shokin   Institute of Computational Technologies SB RAS,
                            Russia
Zhassulan Shaimardanov   D. Serikbayev East Kazakhstan State Technical
                            University, Kazakhstan
Bakytzhan Zhumagulov   National Engineering Academy of RK, Kazakhstan

Program Committee

Maksat Kalimoldayev   Institute of Information and Computing Technologies,
                            Kazakhstan
Galimkair Mutanov   al-Farabi Kazakh National University, Kazakhstan
Nurlan Temirbekov   Kazakhstan Engineering Technological University,
                            Kazakhstan
Amanbek Jaynakov   I. Razzakov Kyrgyz State Technical University,
                            Kyrgyzstan
Igor Bychkov   Institute of System Dynamics and Control Theory
                            of SB RAS, Russia
Viktor Soifer   Korolev Samara State Aerospace University, Russia
Alexander Stempkovsky   Institute for Design Problems in Microelectronics RAS,
                            Russia
Gradimir Milovanovic   Mathematical Institute of SASA, Serbia
Ualikhan Abdibekov   International Kazakh-Turkish University named after
                            Akhmet Yasawi, Kazakhstan
Mikhail Fedoruk   Novosibirsk State University, Russia
Anatoly Fedotov   Institute of Computational Technologies SB RAS,
                            Russia
Sergey Kabanikhin   Institute of Computational Mathematics
                            and Mathematical Geophysics SB RAS, Russia
Vladimir Shaidurov   Institute of Computational Modeling SB RAS, Russia
Sergey Smagin   Computer Center of the Far Eastern Branch RAS,
                            Russia
Igor Bessmertny   ITMO University, Russia
Waldemar Wójcik   Lublin Technical University, Poland
Samir Rustamov   Institute of Management Systems NANA, Azerbaijan

Pavel Vengerik	Lublin Technical University, Poland
Iuri Krak	Taras Shevchenko Kyiv National University, Ukraine
Janusz Partyka	Lublin Technical University, Poland
Györök György	Óbuda University, Hungary
Bo Einarsson	Linköpings Universitet, Sweden
Sergey Bautin	Ural State Transport University, Russia
Egon Krause	Rhine-Westphalian Technical University of Aachen, Germany
Sergey Cherny	Institute of Computational Technologies SB RAS, Russia
Givi Peyman	University of Pittsburgh, USA
Wagdi George Habashi	McGill University, Canada
Andreas Griewank	Humboldt University of Berlin, Germany
Matthias Meinke	Rhine-Westphalian Technical University of Aachen, Germany
Hranislav Milosevic	University of Pristina, Serbia
Vladimir Moskvichev	Special Design Bureau Nauka ICT SB RAS, Russia
Vadim Potapov	Institute of Computational Technologies SB RAS, Russia
Oleg Potaturkin	Institute of Automation and Electrometry SB RAS, Russia
Michael Rasch	High-Performance Computing Center, Stuttgart, Germany
Karl Roesner	Technical University of Darmstadt, Germany
Boris Ryabko	Institute of Computational Technologies SB RAS, Russia
Vladimir Sadovsky	Institute of Computational Modeling SB RAS, Russia
Wolfgang Schroder	Rhine-Westphalian Technical University of Aachen, Germany
Sergei Turitsyn	Aston University, UK
Renjun Wang	Dalian Technological University, China
Ziyaviddin Yuldashev	Mirzo Ulugbek National University of Uzbekistan, Uzbekistan
Yuri Zakharov	Kemerovo State University, Russia
Darkhan Akhmed-Zaki	al-Farabi Kazakh National University, Kazakhstan
Thomas Bönisch	High-Performance Computing Center, Stuttgart, Germany
Denis Esipov	Institute of Computational Technologies SB RAS, Russia
Denis Dutykh	University of Savoy, France
Nina Shokina	University of Freiburg, Germany
Andrey Yurchenko	Institute of Computational Technologies SB RAS, Russia
Oleg Zhizhimov	Institute of Computational Technologies SB RAS, Russia

Organizing Committee

Denis Esipov (Secretary)	Institute of Computational Technologies SB RAS, Russia
Alexey Redyuk	Institute of Computational Technologies SB RAS, Russia
Sergey Rylov	Institute of Computational Technologies SB RAS, Russia
Oleg Sidelnikov	Institute of Computational Technologies SB RAS, Russia
Oleg Gavrilenko	D. Serikbayev East Kazakhstan State Technical University, Kazakhstan
Natalya Denissova	D. Serikbayev East Kazakhstan State Technical University, Kazakhstan
Nazgul Yerdybayeva	D. Serikbayev East Kazakhstan State Technical University, Kazakhstan

Contents

Segmentation Algorithm for Surface Reconstruction According to Data Provided by Laser-Based Scan Point

D. Alontseva[✉], A. Krasavin, A. Kadyroldina, and A. Kussaiyn-Murat

Department of Instrument Engineering and Technology Process Automation,
D. Serikbayev East Kazakhstan State Technical University,
Protozanov. 69, 070004 Ust-Kamenogorsk, Kazakhstan
dalontseva@mail.ru, alexanderkrasavin@mail.ru, akadyroldina@gmail.com,
asselkussaiynmurat@gmail.com

Abstract. The paper presents the results of the elaboration of algorithms of image segmentation and segmentation of the surface based
on the calculation of the local geometric properties of the surface (the
method of the segmentation of the point cloud obtained at the stage of
rough scanning of the surface). The problem of reconstructing a surface
from a spontaneous point cloud has been solved to create a CAD model
based on laser based scan data of the object. The development of the
method of automatic reconstruction of accurate and piecewise smooth
surfaces from spontaneous 3D-points were carried out for designing an
automatic system of path planning for an industrial robot manipulator.
The proposed procedure of automatic segmentation is based on the local
analysis of the Gaussian K and mean H curvatures, obtained by applying
a non-parametric analytical model.

Keywords: Industrial robot manipulator · Point cloud ·
Surface segmentation · 3D-model

1 Introduction

Currently, the use of robots in technological processes has become a model of
high-quality industrial automation throughout the world [1]. The robot is a reliable and effective tool for solving various production tasks: loading, packaging,
plasma welding, cutting, coating, etc. The technologies of plasma spraying of
coatings, as well as the technology of plasma cutting products require accurate
observing a number of technological parameters (the distance from the plasma
system nozzle to the surface of a workpiece, the nozzle movement speed, etc.)
during the entire processing time. Exceeding these parameters beyond the permissible limits can lead not only to defective products, but also to an accident (a
short circuit). In cases when the robot program is generated according to a given
geometrical model of a finished components or part, the deflection of the shape

Y. Shokin and Z. Shaimardanov (Eds.): CITech 2018, CCIS 998, pp. 1–10, 2019.
https://doi.org/10.1007/978-3-030-12203-4_1

of the real object from the model often leads to the violation of technological parameters of processing leading to undesirable consequences. This problem is particularly acute in the case of large-size objects, when small relative errors of geometric parameters and object positioning correspond to unacceptably high absolute deviations of the distances between the tools mounted on the manipulator and the object surface. The deeper solution of these problems is pre-scanning the surface of an object.

A modern robot manipulator can be considered as a means of precise position and orientation of an arbitrary instrument. If a distance sensor or a computer vision system element (camera or projector) is used as a tool, the robot manipulator can be an excellent basis for creating a surface scanning system. It should be noted that the tasks associated with the construction of 3D-scanning systems based on robotic manipulators are of considerable independent interest, both practical and theoretical, and the research stimulated by these tasks is carried out both by research units of companies specializing in the manufacture of test and measurement means and means of automation of production, and representatives of the academic environment [2–9].

The motivation for the elaboration of our robotic scanning system was the analysis of the challenges that we faced when using the industrial robot Kawasaki for plasma surface treatment [10–13], as well as the desire to expand the range of tasks solved by the industrial robots usage. Firstly, we faced the need to applying coatings from biocompatible materials (titanium and alloys based on it, powders of hydroxyapatite, etc.) on the surface of medical products of complex shape. Secondly, the need to cut metal sheets by plasma cutting for small-scale production of large-sized products arises.

The task of reconstructing a surface from a spontaneous point cloud often arises in different engineering applications. We faced the problem of creating a CAD model based on laser-based scanning data of an object, while designing an automatic system of path planning for a robot manipulator. The basic idea of the research is to plan the trajectory based on the scanning data of processed object (finished component).

The main objective of the work is to create a method for automatic reconstruction of accurate and piecewise smooth surfaces from spontaneous 3D points. The segmentation of a point cloud is an important part of constructing a parametric model of an object. The aim of segmentation is partition of an unstructured point cloud into non-intersecting subsets of points. All points within each segment must be close to each other in the selected feature space. The procedure of automatic segmentation proposed by the authors is based on the local analysis of the Gaussian K and mean H curvatures, obtained by applying a non-parametric analytical model.

2   Experimental

An experimental study was conducted with the use of RS-010LA Kawasaki industrial robot (Kawasaki Robotics, Japan). The industrial robot manipulator is a device consisting of movable parts with six degrees of freedom to move to a predetermined profile, managed by a programmable controller E40F-A001. The robot's arm is equipped either by UPR device (NPPTekhnotron Ltd., Russia) for plasma cutting (Fig. 1), or a MPN-004- microplasmatron (E.O. Paton Institute of Electric Welding, Ukraine) for microplasma spraying powders or wires (Fig. 2). Kawasaki robot is controlled by AS software system. Binocular triangulation laser sensors intended for use in automation systems and non-contact measurement of various geometric parameters are chosen as distance sensors.

Fig. 1. The robot's arm equipped by UPR device for plasma cutting of metals

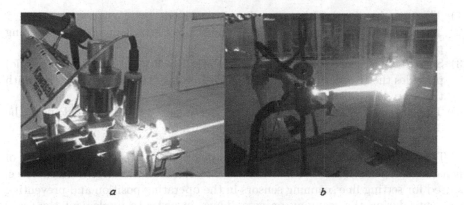

Fig. 2. Robotic processes of microplasma spraying of powder (a) or wire (b)

The KawasakiRS-010LA robot manipulator characteristics are as follows:

- positioning accuracy - 0.06 mm;
- maximal linear speed - 13100 mm/s;
- engagement zone - 1925 mm;
- working load capacity - 10 kg.

In this study, we focused on describing the development of a 3D-scanning system based on proximity sensors mounted on an industrial robot manipulator. The proposed system can scan an object on the basis of the distance from the surface of an object in a discrete set of points, forming a network with the specified geometric parameters imposed on the surface of the object. A 3D-model of the scanned object is built when using interpolation procedures of grid points. Thus, the elaboration of a system of 3D-scanning requires the generation of a data acquisition system involved in job stacking for the robot manipulator. The system consists in stepping according to scanning points, at each point performing actions, followed by the accumulating the results in the storage system. As precision distance sensors usually have a limited operation range, and some of them require orientation of the axis of the sensor in a direction perpendicular to the area of an object subjected to scanning, it is necessary to elaborate a full-blown system consisting of rough and fine scanning subsystems.

3   Results and Discussion

We have developed a 3D-scanning system, where the process of generating a "point cloud" technically performs the following operations:

(1) The scanning route with stop points is specified;
(2) The distance to the surface of the scanned object is measured at the stopping point;
(3) Scan data is sent to a PC. Communication protocol (data transfer logic) provides the transfer of the current coordinates of the scanner together with the scan data (distance);
(4) The obtained data allow us to construct a point cloud and calculate the 3D-coordinates of a set of points on the surface.

The system of rough scanning that uses position sensors with a wide range of measurable distances and is insensitive to the orientation relative to the surface, is used for setting fine scanning sensors in the operating position and preventing accidents during the scanning process. Thus, in order to implement the scanning system we must develop the hardware and software for the data acquisition system briefly described above. The elaboration of the system should include research on testing the algorithms of distance sensors setting in run position and the study of the system specifications in test situations with known geometry. The process of 3D-models creating describes following programming layer of this solution, which encloses kernel software mentioned in previous chapter.

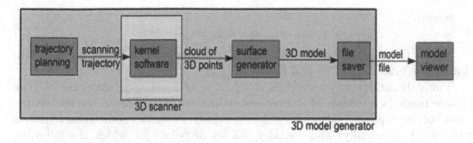

Fig. 3. Blocks of solutions of a higher programming layer

This layer consists in Trajectory Planning Block, Surface Generator Block and File Saver. Placing these blocks in the entire system is shown in Fig. 3 and its functions are described later in the text.

As shown in Fig. 3, output data from kernel 3D-scanner are in the form of 3D-point cloud. It means that spontaneous set of three-dimensional data is produced and sent into the Surface Generator block, where smooth shaded surface should be generated. As it is known, laser-based scanning is characterized by an automatic method for detecting a point cloud, and at the subsequent stages of processing, human intervention is already required. For 3D-models constructing we have to develop a method of segmentation of the point cloud obtained at the stage of rough scanning of the surface, namely the algorithms of image segmentation and segmentation of the surface based on the calculation of the local geometric properties of the surface. Segmentation splits a complex surface into geometrically homogeneous regions with a simple analytical description.

For surface segmenting we suggest using the variation of the region growing algorithm, called Unseeded Region Growing Algorithm [14]. Although this algorithm was proposed for the segmentation of images, its sufficient commonality allows using it for our purposes. The proposed procedure for the automatic segmentation of the surface is based on local analysis of the Gaussian and mean curvature of the surface obtained when constructing a nonparametric analytical model. In the proposed method, the point cloud is interpreted as a height map, i.e. it is assumed that the discrete function of two variables Z (X, Y) on a grid in the plane XY is preassigned.

In each point, the surface is modeled by a second-order polynomial by the local coordinates. The calculation of the coefficients of the approximating polynomial is carried out considering a selected number of neighboring points lying within a circle of a given radius, for the application of the method of least squares. The main advantage of the nonparametric approach is its commonality: previous assumptions about the local geometry of the surface are not introduced, and, moreover, the analytical model of the surface is not proposed.

Let us consider the approximation of the surface in the local neighborhood of an arbitrary point P with the coordinates (x_k, y_m, z_{km}) of type (1)

$$z_{ij} = c_0 + c_1 \cdot (u) + c_2 \cdot (v) + \frac{1}{2} \cdot c_3 \cdot u^2 + c_4 \cdot u \cdot v + \frac{1}{2} \cdot c_{5 \cdot v^2} + \varepsilon_{ij} \qquad (1)$$

where,

$$c_0 = z_{km}, u = x_i - xk, v = y_i - ym \tag{2}$$

Such approximation can be considered as a truncation of Taylor series expansion of the function Z(u,v) at a point (x_k, y_m).

The calculations of coefficients $c_0, c_1, ..., c_5$ of local approximation of type (1) are made by methods of the regression analysis according to the coordinate values of the points neighboring P. More precisely, we specify some value of radius r (selected empirically) and consider the set of points for which the following inequality is satisfied (3)

$$d_{ij} \leq r \tag{3}$$

where d_{ij} is the distance between the point with the coordinates (x_i, y_j, z_{ij}) and point P, which is defined according to (4)

$$d_{ij} = \sqrt{(x_k - x_i)^2 + (y_m - y_j)^2} \tag{4}$$

The calculation of the average and Gaussian curvature of the surface is given below. Let r(u,v) parametric surface. The first fundamental form of surface I can be represented as symmetric matrix (5)

$$I(u, v) = (u^T) \cdot \begin{pmatrix} E & F \\ F & G \end{pmatrix} (v) \tag{5}$$

The first fundamental form completely describes the metric properties of a surface. For example, line element dr can be expressed in terms of the coefficients of the first fundamental form as (6)

$$dr^2 = E \cdot du^2 + 2 \cdot F \cdot du \cdot dv + G \cdot dv^2 \tag{6}$$

The coefficients of the first quadratic form are calculated using the coefficients $c_0, c_1, ..., c_5$ of the approximating polynomial (1) by formulas (7)–(9)

$$E = c_3 / \sqrt{1 + c_1^2 + c_1^2} \tag{7}$$

$$F = c_4 / \sqrt{1 + c_1^2 + c_1^2} \tag{8}$$

$$G = c_5 / \sqrt{1 + c_1^2 + c_1^2} \tag{9}$$

We will denote the partial derivatives of r with respect to u and v by r_u and r_v. Then vector m defined by (10) is a nonzero vector normal to the surface.

$$m = \frac{[r_x, r_y]}{|[r_x, r_y]|} \tag{10}$$

The second fundamental form is a symmetric bilinear form $\prod$ determined by (11)

$$r_{jk} = \frac{\partial^2 r}{\partial x_j \partial x_k} \tag{11}$$

The second quadratic form is a symmetric bilinear form $\prod$ determined by (11)

$$\prod (u, v)) = (u^T) \cdot \begin{pmatrix} L & M \\ M & N \end{pmatrix} \cdot (v) \tag{12}$$

Coefficients of second fundamental form, defined by equations (13)–(15) are the functions of coefficients c_1, c_2

$$L = (r_{11}, m) = 1 + c_1^2 \tag{13}$$

$$M = (r_{12}, m) = c_1 \cdot c_2 \tag{14}$$

$$N = (r_{22}, m) = 1 + c_2^2 \tag{15}$$

The fundamental role of the second quadratic form in the definition of the local surface geometry is that it allows determining the curvature of the given curve belonging to the surface. Namely, the curvature of the normal section is determined by the formula (16)

$$k = \frac{\prod (u, v)}{I (v, v)} \tag{16}$$

In the tangent plane, we can choose a basis $e_1 e_2$ in which the forms of I and $\prod$ at the same time are diagonalized. The directions of vectors e_1 and e_2 are called principal directions, and they are unequivocally determined if $k_1 \neq k_2$. The values of k_1 and k_2 of normal curvatures along the principal directions are called the principal curvatures. They are the extreme values for the normal curvatures at the point, which follows from Euler's formula (17), where φ is the angle between vectors e_1 and v.

$$\frac{\prod (v, v)}{I (v, v)} = k_1 \cdot cos^2 \varphi + k_2 \cdot sin^2 \varphi \tag{17}$$

The product of the principal curvatures at the point is called the Gaussian curvature of the surface at this point:

$$K = k_1 \cdot k_2 = \frac{(L \cdot N - M^2)}{(E \cdot G - F^2)} \tag{18}$$

The sum of the main curvatures at the point is called the average curvature of the surface at this point:

$$H = \frac{1}{2} \cdot (k_1 + k_2) \tag{19}$$

Thus, for each point, the four local curvature values K, H, k_1 and k_2 can be automatically obtained as the functions of the coefficients of the approximating polynomial. In addition, such curvatures are invariant with respect to the adopted reference system, which represents a very important property in analyzing the surface shape. With simultaneous analysis of the sign and the values of

K and H, the classification of the whole point cloud is really achievable, the following surface basic types are possible: hyperbolic ($K = 0$, parabolic (($K < 0$), parabolic ($K = 0$, but $H \neq 0$), flat ($K = H = 0$) and elliptic ($K > 0$) [15]. In addition, the method of empirical optimization of the Taylor expansion bandwidth is presented. This makes it possible to calculate the minimum values of K and H, which can be confirmed by an F-criterion, once the error value of the first kind is fixed. After the 3D-models are created by using the point cloud, they are saved for later viewing.

The analysis of publications [16–18] related to the development of systems for 3D-scanning based on robot manipulators, and the analysis of solutions presented in the market, it is possible to come to a definite conclusion: the systems of 3D-scanning, where robotic manipulators are applied, use either laser triangulation scanners [18], or optical scanners that are implemented on the basis of machine vision systems (in particular, the system using structured light technology). Currently, a number of manufacturers of optical measuring equipment produce 3D-scanners specially designed for systems using a robot manipulator to move the scanner in space. An analysis of the market suggests that they all use methods of machine vision for scanning, i.e., include multiple cameras, and software allowing for the reconstruction of 3D-scenes, located in the working area of the camera system. As an example, let us note the series of MetraSCAN 3D-optical 3D-scanners produced by Creaform Inc. Machine vision systems require much less time for scanning than point scan systems, that require manipulator passing all the scanning points; in addition, point scanning systems are of little use for scanning objects, the surface of which cannot be described by a smooth function – e.g., polyhedra. On the other hand, the point scanning systems of the same accuracy will always be much cheaper and easier to setup and maintain than the corresponding machine vision system. For any optical scanner, the necessary conditions for achieving high measurement accuracy are a high resolution image sensor and a high quality optical system. Machine vision systems require additional expensive hardware and specific software. At the same time, distance sensors for building a fairly accurate scanning system are cheap and easy to use.

Thus, we propose to generate a robot program by a 3D-model of the object to be processed, obtained in the result of scanning, using an industrial robot as a key component of the scanning system. According to the segmentation results, a set of reference points is selected on the surface; and if we know their spatial coordinates, we will be able to fairly accurately construct the 3D model of the object. After selecting reference points, the software generates the program of the manipulator, accomplishing which the manipulator will successively pass the reference points, performing surface scanning at each of the points.

Thus, this paper summarizes principles of three-dimensional scanning system based on robotic manipulator and laser scanner and describes designing of such device. Solution is universal and does not dependent on actually used devices. This approach brings more flexibility on scanned object; almost any type of object can be scanned. Thus, we are developing an intelligent automated system of controlling an industrial robot manipulator, which allows moving a robot arm along a given 3D-trajectory, a model of the product that the robot will coat with plasma. A distinctive feature of the proposed system is pre-3D-scanning the surface of the rough components or the components being processed.

4   Conclusion

Some specific technical solutions and algorithms for developing an intelligent automated system to control an industrial robot manipulator while plasma cutting and processing of large-size products of complex shapes have been suggested, namely: the algorithms of image segmentation and segmentation of the surface based on the calculation of the local geometric properties of the surface (the method of the segmentation of the point cloud obtained at the stage of rough scanning of the surface) and the multi-view 3D-Reconstruction algorithm to quickly scan an object. We are planning to implement automatic generation of program code of a robot-manipulator, taking into account the data of the 3D-scanning of an object to be processed, previously held by means of distance sensors mounted on the robot manipulator. This will allow using components varying in a wide range of geometric parameters and processing products, whose geometrical parameters are determined with low accuracy or products with deviations from a predetermined shape.

Acknowledgment. The study has been conducted with financial support of the Science Committee of the Ministry of Education and Science of the Republic of Kazakhstan within the project AP05130525 "The intelligent robotic system for plasma processing and cutting of large-size products of complex shape".

References

1. Horton, R.: The Robots Are Coming. Deloitte, London (2015)
2. Stumm, S., Neu, P., Brell-Cokcan, S.: Towards cloud informed robotics proceedings. In: Proceedings of the 34th International Symposium on Automation and robotics in Construction, Taipei, Taiwan, pp. 59–64 (2017)
3. Sung, C., Lee, S.H., Kwon, Y.M., Kim, P.Y.: Fast and robust 3D-terrain surface reconstruction of construction site using stereo camera. In: Proceedings of the 33rd International Symposium on Automation and robotics in Construction, Auburn, AL, USA, pp. 19–27 (2016)
4. Chromy, A.: Application of high-resolution 3D-scanning in medical volumetry. INTL J. Electron. Telecommun. **62**(1), 23–31 (2016)
5. Chen, H.M., Chang, K.C.: A cloud-based system framework for storage analysis on big data of massive BIMs. In: Proceedings of the 32nd International Symposium on Automation and Robotics in Construction, Oulu, Finland, pp. 1–8 (2015)
6. Berger, M., et al.: A survey of surface reconstruction from point clouds. In: Proceedings of the Computer Graphics Forum, pp. 1–27. Wiley (2016)
7. Grilli, E., Menna, F., Remondino, F.: A review of point clouds segmentation and classification algorithms. In: The International Archives of the Photogrammetry, Remote Sensing and Spatial Information Sciences, vol. XLII-2/W3, pp. 339–344 (2017)

8. Di Angelo, L., Di Stefano, P.: Geometric segmentation of 3D-scanned surfaces. Comput.-Aided Des. **62**, 44–56 (2015)
9. Li, L., Fan, Y., Zhu, H., Li, D., Li, Y., Tang, L.: An improved RANSAC for 3D-point cloud plane segmentation based on normal distribution transformation cells. Remote Sens. **9**(5), 433–446 (2017)
10. Alontseva, D.L., et al.: Development of the robotic microplasma spraying technology for applying biocompatible coatings in the manufacture of medical products. In: Proceedings AIS 2017–12th International Symposium on Applied Informatics and Related Areas, Székesfehérvár, Hungary, pp. 45–48 (2017)
11. Alontseva, D., Krasavin, A., Prokhorenkova, N., Kolesnikova, T.: Plasma - assisted automated precision deposition of powder coating multifunctional systems. Acta Phys. Pol. A **132**(2), 233–235 (2017)
12. Alontseva, D., Krasavin, A., Nurekenov, D., Ospanov, O., Kusaiyn-Murat, A., Zhanuzakov, E.: Software development for a new robotic technology of microplasma spraying of powder coatings. Przeglad Elektrotechniczny **94**(7), 26–29 (2018)
13. Alontseva, D.L., et al.: Development of technology of microplasma spraying for the application of biocompatible coatings in the manufacture of medical implants. Przeglad Elektrotechniczny **94**(7), 94–97 (2018)
14. Kolmogorov, V., Zabih, R.: Computing visual correspondence with occlusions using graph cuts. In: Proceedings of the 8th International Conference on Computer Vision, ICCV, vol. 2, pp. 508–515 (2001)
15. Crosilla, F., Visintini, D., Sepic, F.: Reliable automatic classification and segmentation of laser point clouds by statistical analysis of surface curvature values. Appl. Geomat. **1**, 17–30 (2009). https://doi.org/10.1007/s12518-009-0002-4
16. Gu, P., Li, Y.: Free-form surface inspection techniques state of the art review. Comput.-Aided Design **36**, 36–48 (2004)
17. Sansoni, G., Trebeschi, M., Docchio, F.: State-of-the-art and applications of 3D imaging sensors in industry, cultural heritage, medicine, and criminal investigation. Sensors **9**, 568–601 (2009)
18. Brosed, F.J., Santolaria, J., Aguilar, J.J., Guillomia, D.: Laser triangulation sensor and six axes anthropomorphic robot manipulator modelling for the measurement of complex geometry products. Robot. Comput.-Integr. Manuf. **28**, 660–671 (2012)

Mathematical Modeling of Temperature Fields in Two-Layer Heat Absorbers for the Development of Robotic Technology for Microplasma Spraying of Biocompatible Coatings

D. L. Alontseva$^{(\boxtimes)}$, A. L. Krasavin, D. M. Nurekenov, and Ye. T. Zhanuzakov

D. Serikbayev East Kazakhstan State Technical University,
Ust-Kamenogorsk, Kazakhstan
dalontseva@mail.ru, alexanderkrasavin@mail.ru, zhan_erzhan@mail.ru
http://www.ektu.kz

Abstract. The motivation for the research was the challenges faced in developing the robotic microplasma spraying technology for applying coatings from biocompatible materials onto medical implants of complex shape. Our task is to provide microplasmatron movement according to the complex trajectory during the surface treatment by microplasma and to solve the problem of choosing the speed of the microplasmatron movement, so as not to cause melting of the coating. The aim of this work was to elaborate mathematical modeling of temperature fields in two-layer heat absorbers: coating-substrate depending on the velocity of microplasmatron with a constant power density. A mathematical model has been developed for the distribution of temperature in two-layer absorbers when heated by a moving source and the heat equation with nonlinear coefficients has been solved by numerical methods.

Keywords: Heat equation · Two-layer metal heat absorbers · Computer simulation

1 Introduction

The multi-purpose methods of Thermal Coating have recently become popular all over the world [1,2]. One of the major methods of gas-thermal deposition of coatings is plasma spraying. The micro plasma spraying (MPS) method is characterized by a small diameter of a spraying spot (1 ... 8 mm) and low (up to 2 kW) power of plasma, which results in low flow of heat into the substrate [3–5]. These characteristics are very attractive for the deposition of coatings with high accuracy, in particular for applying biocompatible coatings in the manufacture of medical implants.

However, the treatment of surfaces of complex shape can be difficult for the implementation of the thermal spraying technology and requires automated

© Springer Nature Switzerland AG 2019
Y. Shokin and Z. Shaimardanov (Eds.): CITech 2018, CCIS 998, pp. 11–22, 2019.
https://doi.org/10.1007/978-3-030-12203-4_2

manipulations of the plasma source and/or the substrate along with robotic control for appropriate surface treatment [1,2].

At present, robot manipulators are widely used in metallurgical industry, automotive industry and mechanical engineering, allowing automating the plasma processing. However, they are used only for large-scale production, because every transition to a new product requires complex calibration procedures to achieve compliance with the model set in the robot previously. Thus, the problem of automatic code generation of a robot program for the model specified by means of CAD is in the limelight of researchers and developers of robotic systems [6–8].

The main prerequisites for the development of the research were the analysis of technical difficulties arising from the industrial robot exploitation for coating by plasma jets, and the desire to expand the scope of tasks solved by the exploitation of an industrial robot. The authors of this paper have carried out a work in the field of application of automated plasma methods of biocompatible or protective coating deposition, described in our paper [9] and protected by certificates of intellectual property [10,11]. One of the main challenges of the robotic technology of microplasma spraying is the choice of modifying irradiation with a microplasma jet in order to set a certain speed of movement of a robot manipulator with a plasma source along the treated surface. Successful deposition of biocompatible coatings with sustained characteristics on parts of complex shape, which are endoprostheses, requires steady travelling with specified speed and power density of the plasma source along the sprayed surface of the product. In order to choose the desired modes of microplasma surface treatment, we need assumptions about the temperature fields that form during irradiation, because it is the temperature that determines the phase transitions in the coatings. The purpose of this study is to develop a mathematical model for the distribution of temperature fields in two-layer absorbers (coating/substrate) under modifying microplasma irradiation of coatings from biocompatible metals (Titanium or Tantalum).

2   Results and Discussion

2.1   Brief Description of the Developed Mathematical Model

We have considered the problem of heating a sample, which is a metal plate (substrate) with the deposited on its surface by a moving axisymmetric heat source coating layer (Fig. 1). The choice of material and layer thicknesses of the absorbers were based on the previously developed scheme of the structure of the protective powder coating applied by a plasma jet on the steel substrate described in [12,13].

The task of heating a plate by a moving flat heat source comes down to the solution of a boundary value problem for a differential equation of heat conductivity. As thermal characteristics of metals, such as thermal conductivity and specific heat, highly depend on temperature, and in processes of radiation treatment of coatings, temperature difference between various points of the sample

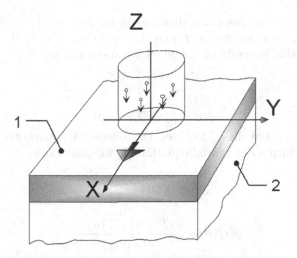

Fig. 1. Moving flat heat source, where 1-coat; 2-substrate; XYZ - moving Cartesian coordinate system.

can exceed 1000K (flash-off of a surface of the sample at maintaining the temperature of end faces of a plate close to room), adequate mathematical model of thermal processes at radiation treatment of coatings is a non-linear heat conduction equation considering dependence of thermal characteristics of material on temperature.

2.2 Non-linear Heat Conduction Equation. Kirchhoff Transformation

The heat conduction equation for the homogeneous environment whose thermal characteristics depend on temperature in case of lack of sources of heat distributed in the environment, is:

$$div(k(T)\nabla(T)) = c(T)\rho(T)\frac{\partial T}{\partial t}. \tag{1}$$

In the Eq. (1) T (x, y, z, t) a dynamic field of temperatures taken in an absolute scale (degrees Kelvin), k(T) - function of dependence of a thermal conductivity of substance on temperature, c(T) function of dependence of specific heat of substance on temperature, $\rho(T)$ - function of dependence of density of substance on temperature. Further on, we will use a differential equation (5) that the Eq. (1) turns into after Kirchhoff transformation [14].

$$\vartheta(x, y, z, t) = \int_{T_0}^{T(x,y,z,t)} k(\tau)d\tau \tag{2}$$

For the sake of convenience, we believe value of the T_0 parameter equal to environment temperature. As at any moment t temperature $T(P,t)$ in the arbitrariest point P meets condition $T(P,t) \geq T_0$, taking into account nonnegativity of the function k(T) we can claim that for display $T(x, y, z, t) \to \vartheta(x, y, z, t)$, set by formula (2) exists the inverse display, i.e. there is a $T = T(\vartheta)$.

The remarkable property of Kirchhoff's transformation that

$$\nabla \vartheta = k(T) \cdot \nabla T \tag{3}$$

and, as a result

$$div(k(T) \cdot \nabla T) = \Delta \vartheta \tag{4}$$

Taking into account these identities, the differential equation of heat conductivity (1) turns into a differential equation (5) for function ϑ

$$\Delta \vartheta = Q(\vartheta) \cdot \frac{\partial \vartheta}{\partial t} \tag{5}$$

where

$$Q(\vartheta) = \frac{c(T(\vartheta)) \cdot \rho(T(\vartheta))}{k(T(\vartheta))} \tag{6}$$

To apply transformation of Kirchhoff to the solution of boundary value problems of the theory of heat conductivity, it is necessary to reformulate the boundary conditions set for function T, in boundary conditions for function ϑ. Further on we will believe that the boundary value problem is formulated for the given area of space Ω, and a symbol $\partial\Omega$ will be designated as an area border. Let's consider separately cases of boundary conditions of the 1st and 2nd sort.

(A) The boundary conditions of the 1st sort set by the Eq. (7)

$$\vartheta_{in}(M, t) = \vartheta(T_{in}(M, t)) \tag{7}$$

where $T_{in}(M, t)$ the function setting distribution of temperatures on area border. As it has been shown above, there is a uniquely determinated function $\vartheta(T)$. Let's define on area border Ω function ϑ_{in}

$$\vartheta_{in}(M, t) = \vartheta(T_{in}(M, t)) \tag{8}$$

Then a boundary condition (6) for function $T(x, y, z, t)$ will l turn into a boundary condition (9) for function $\vartheta(x, y, z, t)$.

$$\vartheta_{|\partial\Omega} = \vartheta_{in}(M, t) \tag{9}$$

(B) The boundary conditions of the 2nd sort set by the Eq. (10)

$$- (k(T) \cdot \frac{\partial T}{\partial \overrightarrow{n}})_{\partial\Omega} = P(M, t) \tag{10}$$

where function $P(M, t)$ describes the surface density of power of the heat sources affecting the area border. Owing to equality (2), the boundary condition for function ϑ looks like (11).

$$-(\frac{\partial \vartheta}{\partial \overrightarrow{n}})_{\partial\Omega} = P(M, t) \tag{11}$$

The Eq. (5) in comparison with the Eq. (1) has much simpler structure allowing using well developed potential theory methods for the decision in some cases.

2.3   The Limiting Stationary State When Heating a Plate by Moving a Flat Heat Source. Differential Equation of the Limiting Steady-State

Let's introduce a concept of the limiting steady-state when heating a body by moving heat source. For this purpose we will consider the task given below about heating the semi-infinite body by moving heat source. The obtained results hereafter naturally extend multilayer plates of the terminating sizes

Let axes X and Y of a Cartesian coordinate system lie on the surface of the homogeneous semi-infinite body, with thermal characteristics of $k(T)$, $c(T)$, and density $\rho(T)$. Axis Z is sent to the body depth (at such choice of a frame, the semi-infinite body represents area of space set by inequality $z \geq 0$). Let's assume that in an initial instant of $t = 0$ the flat source of heat moving with constant speed $\vec{v} = v \cdot \vec{i}$, where v - the speed module, and $\vec{i}$ - a coordinate basis vector of collinear axes X begins to act on a surface of sample. Let us assume that the source of heat has an axial symmetry, and in an initial instant axis Z of a Cartesian coordinate system coincides with a source axis. Let us note that the assumption of a axial symmetry of the heat source is insignificant at a conclusion of a differential equation of the limiting steady-state, and it is introduced, first, for simplification, secondly, as the case that is most often found in practice. Let the surface power density of a source be described by the $P(r)$ function where r distance to a source. In that case, a dynamic temperature profile of $T(x, y, z, t)$ in a semi-infinite body will satisfy a differential equation (1) and a regional condition (12) on the sample surface (z=0 plane), the initial conditions (13) and a condition (14)

$$- (k(T) \cdot \frac{\partial T}{\partial \vec{n}})_{z=0} = P(\sqrt{(x - \vartheta \cdot t)^2 + y^2 + z^2}) \tag{12}$$

$$T(x, y, z, 0) = T_0 \tag{13}$$

$$\forall x, y, t \lim_{z \to \infty} T(x, y, z, t) = T_0 \tag{14}$$

Let us turn into the relative frame Cartesian coordinate system moving with a speed of v, with axes of X^*, Y^*, Z^* parallel to axes X, Y, Z of the above described fixed frame. Coordinates of a point x^*, y^*, z^* in a relative frame of logical coordinates are connected with coordinates x, y, z of the same point in the fixed frame (15), (16) and (17).

$$x = x^* + \vartheta \cdot t \tag{15}$$

$$y = y^* \tag{16}$$

$$z = z^* \tag{17}$$

Let $T^*(x^*, y^*, z^*, t)$ the function describing a temperature field in a relative frame of coordinates. Using differentiation rules and coordinate transformation formulas (15), (16) and (17) we will obtain

$$\frac{\partial T}{\partial x} = \frac{\partial T^*}{\partial x^*} \tag{18}$$

$$\frac{\partial T}{\partial y} = \frac{\partial T^*}{\partial y^*} \tag{19}$$

$$\frac{\partial T}{\partial z} = \frac{\partial T^*}{\partial z^*} \tag{20}$$

$$\frac{\partial T}{\partial t} = \frac{\partial T^*}{\partial t^*} - \vartheta \frac{\partial T^*}{\partial x^*} \tag{21}$$

Thus, the differential equation (1) turns into a differential equation (22) for the function T^*:

$$\frac{\partial}{\partial x^*}(k \cdot \frac{\partial T^*}{\partial x^*}) + \frac{\partial}{\partial y^*}(k \cdot \frac{\partial T^*}{\partial y^*}) + \frac{\partial}{\partial z^*}(k \cdot \frac{\partial T^*}{\partial z^*}) = c \cdot p \cdot (\frac{\partial T^*}{\partial t} - \vartheta \frac{\partial T^*}{\partial x^*}) \tag{22}$$

The Eq. (22) can be considered a special case of the equation of Fourier Ostrogradsky for the moving environment [9]:

$$div(k \cdot \nabla T) + q_v = p \cdot c \cdot \frac{DT}{dt} \tag{23}$$

In the Eq. (23) q_v - apparent density of sources of heat, and a $\frac{DT}{dt}$ symbol designates the substantive derivative T determined by a formula (24)

$$\frac{DT}{dt} = \frac{\partial T}{\partial t} + (\nabla T, \vec{v}) \tag{24}$$

It is logical to assume that when the travel time of source is aiming to infinity, in the frame traveling together with a source, the quasistationary temperature profile will be observed, in other words, we can put in the Eq. (23) $\frac{\partial T^*}{\partial t^*} = 0$. We will name this mode the limiting steady state, described by differential equation (25).

$$\frac{\partial}{\partial x^*}(k \cdot \frac{\partial T^*}{\partial x^*}) + \frac{\partial}{\partial y^*}(k \cdot \frac{\partial T^*}{\partial y^*}) + \frac{\partial}{\partial z^*}(k \cdot \frac{\partial T^*}{\partial z^*}) + \vartheta \cdot (\frac{c \cdot \rho}{k}) \cdot (k \cdot (\frac{\partial T^*}{\partial x^*})) = 0 \tag{25}$$

Applying transformation of Kirchhoff to this equation, we will obtain

$$\triangle \vartheta + v \cdot Q(\vartheta) \cdot \frac{\partial \vartheta}{\partial x} = 0 \tag{26}$$

2.4   Heating of Semi-infinite Body by Moving Flat Heat Source

This section is devoted to the description of the numerical method of task solution of heating a half-space by a moving heat source. It should be noted that this task solution allows finding rather precise estimates of a temperature schedule in the field of border of a substrate coating.

First of all, we will formulate a boundary value problem for function $\vartheta(x, y, z)$ representing transformation of Kirchhoff of a temperature field $T(x, y, z)$, in a coordinate system. As it has been shown above, function $\vartheta(x, y, z)$ o satisfy a differential equation (23) in the $z \geq 0$

$$\nabla \vartheta + v \cdot Q(\vartheta) \cdot \frac{\partial \vartheta}{\partial x} = 0 \qquad (27)$$

At boundary conditions (28) and (29):

$$\left(\frac{\partial \vartheta}{\partial z}\right)_{z=0} = -P(x, y) \qquad (28)$$

$$\lim_{z \to \infty} \vartheta(x, y, z) = \vartheta_0 \qquad (29)$$

If (29) a constant ϑ_0 - transformation of Kirchhoff of environment temperature T_0.

This boundary value problem can be reduced to a non-linear integral equation for the numerical solution n which would make it possible to develop the iterative method which enters the group of methods of the fixed point of the squeezing operator finding. As well as in many cases of application of iterative methods, in the case considered by us the choice of an initial approximation influences the speed of calculations. For initial approximation finding, we used a method of a linearization of a differential equation (27). It should be noted that in many cases solution of the linearized equation (27) can serve an appropriate approximation of solutions of initial non-linear equation (27).

For further consideration it is handier to use an invariant form of the equation (27)

$$\Delta \vartheta + \nabla \cdot (F(\vartheta) \cdot \overrightarrow{v}) = 0 \qquad (30)$$

where $F(\vartheta)$ is the antiderivative of the function $Q(\vartheta)$

$$\frac{dF(\vartheta)}{d\vartheta} = Q(\vartheta) \qquad (31)$$

Certainly, the $F(\vartheta)$ function is determined within the arbitraries additive constant. For the sake of convenience we will assume

$$F(\vartheta) = \int_{\vartheta_0}^{\vartheta} Q(\vartheta) d\vartheta \qquad (32)$$

To show equivalence of the Eqs. (30) and (27) we will consider expression $\nabla \cdot (F(\vartheta) \cdot \overrightarrow{v})$ (divergence of a field of vectors $F(\vartheta) \cdot \overrightarrow{v}$):

$$\nabla \cdot (F(\vartheta) \cdot \overrightarrow{v}) = (\nabla F, \overrightarrow{v}) + F \cdot (\nabla \cdot \overrightarrow{v}) \qquad (33)$$

as

$$\nabla F(\vartheta) = \frac{dF(\vartheta)}{d\vartheta} \cdot \nabla \vartheta \qquad (34)$$

and divergence of a constant field of vectors $\overrightarrow{v}(x, y, z) = (v, 0, 0)$ equals to zero,
$\nabla \cdot \overrightarrow{v} = 0$

$$\nabla \cdot (F(\vartheta) \cdot \overrightarrow{v}) = \frac{dF(\vartheta)}{d\vartheta} \cdot (\nabla \vartheta, \overrightarrow{v}) \tag{35}$$

Whence it follows that taking into account (31) we obtain

$$\nabla \cdot (F(\vartheta) \cdot \overrightarrow{v}) = v \cdot Q(\vartheta) \cdot \frac{\partial \vartheta}{\partial x} \tag{36}$$

Let's look for a scalar field ϑ in the form of superposition of fields φ and η

$$\vartheta(x, y, z) = \varphi(x, y, z) + \eta(x, y, z) \tag{37}$$

where a scalar field $\varphi(x, y, z)$ satisfies the equation of Laplace (38)

$$\Delta \varphi = 0 \tag{38}$$

and the boundary condition (39):

$$(\frac{\partial \varphi}{\partial z})_{z=0} = P(x, y) \tag{39}$$

The boundary condition (39) forms the boundary value problem 3 for the Laplace equation (38).

Thus, we obtain the following boundary-value problem for the function $\eta(x, y, z)$:

To find function $\eta(x, y, z)$ defined in a half-space $z \geq 0$ and satisfying in it a differential equation (40)

$$\Delta \eta = -\nabla \cdot (F(\varphi + \eta) \cdot \overrightarrow{v}) \tag{40}$$

and boundary condition(41) on border of area:

$$(\frac{\partial \eta}{\partial z})_{z=0} = 0 \tag{41}$$

Below we will show how the boundary value problem given above can be reduced to an integral equation for function $\eta(x, y, z)$.

Let's consider the following boundary value problem for a Poisson equation:
Let function $f(x, y, z)$ satisfies in field of $z \geq 0$ a Poisson equation (42):

$$\Delta f = \rho \tag{42}$$

where $\rho(x, y, z)$ the given function on a border (plane $z = 0$) boundary condition (43):

$$(\frac{\partial f}{\partial z})_{z=0} = 0 \tag{43}$$

As the solution of this boundary value problem serves function (44)

$$f(x, y, z) = \int \int \int G(x', y', z', x, y, z) \cdot \rho(x', y', z') \cdot dx'dy'dz' \qquad (44)$$

Where $G(x', y', z', x, y, z)$ the Green function of this boundary value problem determined by formulas (44) and (45)

$$G(x', y', z', x, y, z) = \frac{1}{4 \cdot \pi} \cdot (\frac{1}{r(x', y', z', x, y, z)} + \frac{1}{r(x', y', z', x, y, -z)}) \qquad (45)$$

$$r(x', y', z', x, y, z) = \sqrt{(x - x')^2 + (y - y')^2 + (z - z')^2} \qquad (46)$$

Note: If we use physical interpretation of a Poisson equation in which function $f(x, y, z)$ describes a stationary temperature field, and the function $\rho(x, y, z)$ is related to the heat density distribution function $p(x, y, z)$ the ratio and the coefficient of thermal conductivity of the environment λ by the relation (47),

$$\rho(x, y, z) = \frac{1}{\lambda} \cdot p(x, y, z) \qquad (47)$$

Then a design of a Green's function of the above described boundary value problem can be considered as natural generalization of a method of images.

Thus, the boundary value problem (1) for a differential equation (47) comes down to a non-linear integral equation

$$\eta(x, y, z) = -\int \int \int (\nabla \cdot (F(\eta + \varphi) \cdot \vec{v})) \cdot G(x', y', z', x, y, z) dx'dy'dz' \qquad (48)$$

In the last equation the nabla operator represents a symbolic vector

$$\nabla = (\frac{\partial}{\partial x'}, \frac{\partial}{\partial y'}, \frac{\partial}{\partial z'}) \qquad (49)$$

And the factor $\nabla \cdot (F(\eta + \varphi) \cdot \vec{v})$ in expanded form registers as

$$\nabla \cdot (F(\eta(x', y', z') + \varphi(x', y', z')) \cdot \vec{v}) \qquad (50)$$

Further we suppose that there is a rectangular parallelepiped Ω, (defined as $M(x, y, z) \ni \Omega$ if $x_{max} \geq x \geq -x_{max}, y_{max} \geq y \geq -y_{max}$ and $0 \geq z \geq z_{max}$, such, that $F \cdot (\eta(x, y, z) + \varphi(x, y, z)) = 0$ at any point $A(x, y, z)$ lying outside this area.

Let's transform a right member of the Eq. (48), integrating piecemeal

$$(\nabla \cdot (F(\eta + \varphi) \cdot \vec{v})) \cdot G = \nabla \cdot ((F(\eta + \varphi) \cdot G)) \cdot \vec{v} - F(\eta + \varphi) \cdot (\nabla G, \vec{v})) \qquad (51)$$

According to the theorem of *Gauss − Ostrogradsky*

$$\int \int \int_\Omega \nabla \cdot ((F(\eta + \varphi) \cdot G) \cdot \vec{v}) dx'dy'dz' = \int \int_D F(\eta + \varphi) \cdot G \cdot (\vec{n}, \vec{v}) dS \qquad (52)$$

where $\vec{n}$ - a vector of a normal to a surface of border D of the area Ω.

On the plane $z = 0$ vector $\vec{n} = (0, 0, 1)$ is orthogonal to $\vec{v}$ vector, i.e., on sides of a parallelepiped D not lying in the plane $z = 0$ $F(\eta + \varphi) = 0$.

Thus,

$$\int \int \int_\Omega \nabla \cdot ((F(\eta + \varphi) \cdot G) \cdot \vec{v}) dx' dy' dz' = 0 \tag{53}$$

and taking into account (50) we obtain

$$\eta(x, y, z) = \int \int \int_\Omega F(\eta + \varphi) \cdot (\nabla G, \vec{v}) dx' dy' dz' \tag{54}$$

The problem of finding the solution of a non-linear integral equation (54) can be interpreted as a problem of finding the fixed point of display $f(x, y, z) \rightarrow Kf(x, y, z)$, where action of the non-linear operator K is defined by expression (55)

$$Kf(x, y, z) = \int \int \int_\Omega F(\eta(x', y', z') + \varphi(x', y', z')) \cdot (\nabla G, \vec{v}) dx' dy' dz' \tag{55}$$

We make a hypothesis, that display (55) is the squeezing display, i.e. we assume that there is a constant $d < 1$, such that $\forall f_1, f_2 \in U$ is carried out inequality:

$$\| Kf_1 - Kf_2 \| < d \cdot \| f_1 - f_2 \| \tag{56}$$

At the same time we designate a symbol U metric space of the square integrable functions defined in the area Ω, with a reference metrics:

$$\| f \| = \int \int \int_\Omega f^2 dx dy dz \tag{57}$$

It is known that for the squeezing operator K, the repetitive process determined by the equations looks like (58)

$$f_k = Kf_{k-1} \tag{58}$$

meets to the fixed point of operator K at any initial approximation of f_0.

3   Conclusion

A mathematical model for the distribution of temperature fields in two-layer absorbers with modifying microplasma irradiation of metallic coatings has been developed and the numerical method for calculating temperature fields in the coating/substrate's system heating by a moving heat source has been designed. Kirchhoff's transformation was applied when solving a nonlinear heat equation by numerical methods. Based on the calculations of the temperature fields, certain speeds of movement of the robot arm with a microplasma source and certain power densities of the plasma source, i.e., microplasma surface treatment modes were recommended in order to ensure the desired temperature distribution in the coating/substrate's system. Coatings from biocompatible materials deposited by the microplasma according to recommended modes onto steel and titanium substrates have been obtained. It is shown that the robotic microplasma spraying method allows applying a wide range of biocompatible materials: Co-based powders, Titanium or Tantalum wires onto medical implants. Thus, the applied value of the developed mathematical model and numerical methods for solving problems of robotic microplasma spraying of biocompatible coatings on endoprostheses has been shown. The results of the research are of significance for a wide range of researchers developing numerical methods for solving nonlinear equations.

Acknowledgment. The study has been conducted with financial support of the Science Committee of the Ministry of Education and Science of the Republic of Kazakhstan within the framework of program-targeted financing for 2017–2019 years on the scientific and technical sub-program 0006/PCF-17 "Manufacture of titanium products for further use in medicine".

References

1. Tucker, R.C. (ed.): Introduction to coating design and processing. In: ASM Handbook: Thermal Spray Technology, vol. 5A, pp. 76–88 (2013)
2. Vardelle, A., Moreau, C., Nickolas, J., Themelis, A.F.: A perspective on plasma spray technology. Plasma Chem. Plasma Process. **35**, 491–509 (2015). https://doi.org/10.1007/s11090-014-9600-y
3. Lugscheider, E., Bobzin, K., Zhao, L., Zwick, J.: Assessment of the microplasma spraying process for coating application. Adv. Eng. Mater. **8**(7), 635–639 (2006). https://doi.org/10.1002/adem.200600054. Special Issue: Thick Coatings for Thermal, Environmental and Wear Protection
4. Borisov, Yu., Sviridova, I., Lugscheider, E., Fisher, A.: Investigation of the microplasma spraying processes. In: The International Thermal Spray Conference, Essen, Germany, pp. 335–338 (2002)
5. Andreev, A.V., Litovchenko, I.Y., Korotaev, A.D., Borisov, D.P.: Thermal stability of Ti-C-Ni-Cr and Ti-C-Ni-Cr-Al-Si nanocomposite coatings. In: 12th International Conference on Gas Discharge Plasmas and Their Applications. IOP Publishing Journal of Physics: Conference Series, vol. 652 (2015). https://doi.org/10.1088/1742-6596/652/1/012057
6. Nelayeva, E.I., Chelnokov, Y.N.: Solution to the problems of direct and inverse kinematics of the robots-manipulators using dual matrices and biquaternions on the example of stanford robot arm. Mechatron. Autom. Control **16**(7), 456–463 (2015)
7. Rodrigues, M., Kormann, M., Schuhler, C., Tomek, P.: Robot trajectory planning using OLP and structured light 3D machine vision. In: Bebis, G., et al. (eds.) ISVC 2013. LNCS, vol. 8034, pp. 244–253. Springer, Heidelberg (2013). https://doi.org/10.1007/978-3-642-41939-3_24

8. Brosed, F.J., Santolaria, J., Aguilar, J.J., Guillomia, D.: Laser triangulation sensor and six axes anthropomorphic robot manipulator modelling for the measurement of complex geometry products. Robot. Comput.-Integr. Manuf. **28**, 660–671 (2012)
9. Alontseva, D., Krasavin, A., Prokhorenkova, N., Kolesnikova, T.: Plasma - assisted automated precision deposition of powder coating multifunctional systems. Acta Phys. Pol. A **132**(2), 233–235 (2017)
10. Krasavin, A.L., Alontseva, D.L., Denissova N.F.: Calculation of temperature profiles in the two-layer absorbers with constant physical characteristics heated by a moving source. Certificate of authorship No. 0010558 of the Republic of Kazakhstan for the computer program, No. 1151 of August 20 (2013)
11. Nurekenov, D.M., Krasavin, A.L., Alontseva, D.L.: Converter for DXF drawings into AS language of robot manipulator Kawasaki RS010L. Certificate of authorship No. 009030 of the Republic of Kazakhstan for the computer program, No. 1490 of June 21 (2017)
12. Alontseva, D., Ghassemieh, E.: The structure-phase compositions of powder Ni-based coatings after modification by DC plasma jet irradiation. J. Phys.: Conf. Ser. **644**, 012009 (2015). https://doi.org/10.1088/1742-6596/644/1/012009. Electron Microscopy and Analysis Group Conference (EMAG 2015)
13. Alontseva, D.: The Chapter 3: Structure and mechanical properties of nanocrystalline metallic plasma-detonation coatings. In: Aliofkhazraei, M. (ed.) Comprehensive Guide for Nanocoatings Technology, vol. 3, pp. 53–84. Nova Science Publishers Inc., New York (2015). ISBN 978-1-63482-647-1
14. Kim, S.: A simple direct estimation of temperature-dependent thermal conductivity with Kirchhoff transformation. Int. Commun. Heat Mass Transf. **28**(4), 537–544 (2001). https://doi.org/10.1016/S0735-1933(01)00257-3

Environmental Threat Calculation Dealing with the Risk of Industrial Atmospheric Emission

A. Baklanov$^{(\boxtimes)}$, T. Dmitrieva, K. Tulendenova, A. Bugubayeva, and M. Rakhimberdinova

D. Serikbaev East Kazakhstan State Technical University, Ust-Kamenogorsk, Kazakhstan
abaklanov@ektu.kz

Abstract. The following work proposes an estimation procedure of risks based on the calculation modeling action of concentration of the environmental exposure thrown into atmosphere by industrial enterprises with consideration for atmosphere perseverance category and wind rose in the region. The information system of risk assessment employing the suggested mathematical model for computation of environmental hazards is described in the article.

The information system has client/server architecture and uses OLAP technology to get the necessary information. The received data on Ust-Kamenogorsk allowed the authors to analyze the weighting risk coefficient effect on per unit value of environmental threat. The calculation is performed for the city of Ust-Kamenogorsk.

Keywords: Ecology · Risks · Damage · Emissions · Air pollution

1 Introduction

On December 12 2015 the Paris Climate Agreement was accepted. 195 forum participants decided to not let the planet average temperature increase more than 2° by 2100 in comparison with the pre-industrial era. However, the document does not provide for quantitative obligations of reducing or limiting CO_2 emissions [1].

Kazakhstan announced its intentions to reduce emissions by 15% and conditionally by 25% with additional international support by 2030 from the base year of 1990 [2]. The set goals can contribute to the low-carbon "green" development path.

The main sources of environment pollution and degradation of natural systems are industry, agriculture, motor vehicles and other anthropogenic factors. Of all the constituent components of the biosphere and the environment, the atmosphere is the most sensitive. It is the first to receive polluting substances not only in gaseous, but also liquid and solid state.

© Springer Nature Switzerland AG 2019
Y. Shokin and Z. Shaimardanov (Eds.): CITech 2018, CCIS 998, pp. 23–33, 2019.
https://doi.org/10.1007/978-3-030-12203-4_3

Man pollutes the atmosphere for thousands of years, however the consequences of the use of fire, which he has used all this period were not significant. What is the atmosphere? The air around us is a mixture of gases or, in other words, the atmosphere enveloping our globe. The afflux of various pollutants into the atmosphere from stationary industrial sources makes currently more than 4 million tons per year.

A significant amount of highly toxic gaseous and solid substances is released into the atmosphere over Kazakhstan. If we compare the number of emissions from various stationary sources, about 50% is emitted by heat and power sources and 33% by mining and non-ferrous metallurgy enterprises. The largest number of emissions of various pollutants takes place in East Kazakhstan-2231.4 thousand tons/year, which males 43% of the total emissions around Kazakhstan.

The second place in the number of emissions belongs to Central Kazakhstan-1868 thousand tons/year or 36%. The least polluted is the atmosphere in Northern Kazakhstan: 363.2 thousand tons/year (7%) and Southern Kazakhstan: 415.1 thousand tons/year, which amounts to 8%. The most mobile, with a wide range of action, are oxides of nitrogen and sulfur. They cover considerable distances and have a strong impact on destruction of crops in the first place.

A significant contribution to the pollution of the air basin and other environmental components is made by motor vehicles of the Republic. Their emissions, especially in urban areas, range from 25 to 50%. According to the pollution of the atmosphere with motor exhaust gases in the first place is Almaty - 75%, then Aktobe - 47.1, Semey 46.6, Zhambyl - 43.1, and Ust-Kamenogorsk - 41.4%. Zhezkazgan has the least content of exhaust gas in the atmosphere - 14.8. Then go Petropavlovsk - 26.3 and Ridder - 27.6%. However, the highest pollutant gas content of the atmospheric air, oddly enough was established in the cities such as Kostanay - 84.7% and Uralsk - 81.7%, where the number of industrial enterprises and vehicles is relatively fewer than in the above-mentioned cities. Motor vehicles are the main pollutants of air and, to a certain extent, of soil and water. According to statistics, there are more than 200 thousand cars per more than a million Almaty population-today.

Pollution of the atmosphere of cities with solid and gaseous pollutants reduces the intensity of sunlight, clogs the air with a significant amount of solid particles, which serve as concentration nuclei, contributing to the emergence of fogs and smogs. The high content of harmful impurities in the atmosphere in the solid and gaseous state affects the thermal properties of the atmosphere. Under the influence of sunlight, as a result of photochemical reactions, a summation effect is formed, thus contributing to the emergence of new, more toxic substances that cause smog.

2 Mathematical Model

The environmental hazard is determined by two factors for the population living in a territory with a high industrial content: damage from actual danger and risk (potential hazard) in the event of emergency situations. For this reason the environmental hazard value in relative terms can be presented in the form [3]:

$$G_{RT} = \phi(Y_{RT}, R_{RT}), \tag{1}$$

where G_{RT} – environmental hazard risk in relative terms; Y_{RT} – damage in dimensionless relative terms; R_{RT} – risk in dimensionless relative terms.

The conditions determined by physical laws can be written in the following form:

$$\frac{d\phi}{dY_{RT}} > 0; \frac{d\phi}{dR_{RT}} > 0; \phi(0, R_{RT} > 0); \phi(Y_{RT} > 0, 0); \phi(0,0) = 0. \tag{2}$$

A similar approach to risk calculation was used in the articles [4,5]. The damage and risk characteristics independence requires the ϕ function presentation in their product form. Considering the conditions 2 the proposed ϕ function form in [3] allows us to write the environmental risk value in the following form:

$$G_{RT} = (Y_{RT} + 1)^{P_Y} \cdot (R_{RT} + 1)^{P_R} - 1. \tag{3}$$

where P_Y and P_R – weighted coefficients which are describe the damage and risk fractional contribution to the environmental hazards value.

According to the model proposed in [3], the damage to the Y_{RT} population consists of two components: direct damage to Y_{DD} immediately inflicted to the population and indirect damage Y_{IND} caused to the population due to habitat degradation:

$$Y_{RT} = (Y_{DD} + 1)^{P_{YDD}} \cdot (Y_{IND} + 1)^{P_{RIND}} - 1. \tag{4}$$

The specific weight coefficients choice of direct P_{YDD} and indirect P_{RIND} damages depends on the natural environment efficiency on the vital activity of the population living in the given territory. In the case of a natural environment marked impact on human living conditions the equivalent factors model can be accepted, where the weight coefficients are equal and have value 0.5. Must be chosen the selected factors model and particularly accept in the case of a natural environment weak impact on human living conditions:

$$P_{YDD} : P_{RIND} = 9 : 1. \tag{5}$$

Considering the industrial emissions impact in the atmosphere we have [3] the following expressions for Y_{DD} and Y_{IND}:

$$Y_{DD} = \sum_{i=1}^{n} \frac{C_i^{emis}}{MPC_i^{emis}} \cdot \frac{N_{ter}}{N_{cntr}}, \tag{6}$$

$$Y_{IND} = \sum_{i=1}^{n} \frac{C_i^{emis}}{MPC_i^{emis}} \cdot \frac{S_{ter}}{S_{reg}} \cdot \beta. \tag{7}$$

where C_i^{emis} – given territory actual (measured) concentration in the atmosphere of the i-th substance;

MPC_i^{emis} – maximum allowable concentration of the i-th substance in the air;

N_{ter} – population density living on the polluted territory under consideration;

N_{cntr} – country average population density;

S_{ter} – polluted atmosphere area;

S_{reg} – ecologically homogeneous region area that includes given territory;

β – given territory significance index in conserving the natural environment in the region $(0 \leq \beta \leq 1)$.

Expressions 4–7 are allowed to calculate $YO\Pi$ - damage in dimensionless relative terms. The general expression that allows estimating the damage numerically appears as follow:

$$Y_{RT} = \left(\sum_{i=1}^{n} \frac{C_i^{emis}}{MPC_i^{emis}} \cdot \frac{N_{ter}}{N_{cntr}} + 1\right)^{0.9} \cdot \left(\sum_{i=1}^{n} \frac{C_i^{emis}}{MPC_i^{emis}} \cdot \frac{S_{ter}}{S_{reg}} \cdot \beta + 1\right)^{0.1} - 1. \quad (8)$$

Similarly, R_{RT} is calculated in dimensionless relative indicators:

$$R_{RT} = \left(\sum_{i=1}^{n} \frac{C_i^{emis}}{MPC_i^{emis}} \cdot \frac{N_{ter}}{N_{cntr}} + 1\right)^{0.9} \cdot \left(\sum_{i=1}^{n} \frac{C_i^{emis}}{MPC_i^{emis}} \cdot \frac{S_{ter}}{S_{reg}} \cdot \beta + 1\right)^{0.1} - 1. \quad (9)$$

The damage and risk weight coefficients can be calculated using formulas that express the relative cost parameters contribution of damage and risk in the environmental hazard total cost:

$$P_Y = \frac{Y_{VT}}{G_{VT}}; P_R = \frac{R_{VT}}{G_{VT}}. \quad (10)$$

where Y_{VT} – general integrated damage to the territory in value term;

R_{VT} – risk cost parameters;

G_{VT} – environmental hazard total cost $Y_{VT} + R_{VT}$.

The following expressions are used based on the proposed damage and risk presentation in value terms:

$$Y_{VT} = Y_{NORM} + Y_{POP}. \quad (11)$$

where Y_{NORM} – damage cost from the environment (normatively determined damage) direct pollution;

Y_{POP} – social and ecological damage cost due to deteriorating the population living conditions.

Expression 11 can be written in the following form:

$$Y_{VT} = a \cdot \sigma \cdot f \cdot M_{COND} + Y_G + Y_Q + Y_W. \quad (12)$$

where a – proportionality cost factor for the conditional air contaminant (tg/cond.t);

σ – dimensionless coefficient, which considers the territory peculiarities;

f – dimensionless coefficient that considers the contaminant fractions size, the dispersal pattern, and the subsidence rate in the atmosphere;

M_{COND} – annual release reduced mass in the atmosphere from conditional polluter (taking into account its ecological danger) (cond.t./year);

Y_G – reduced annual damage, which is connected with decreasing population growth rate parameter;

Y_Q – reduced annual damage that is connected with decreasing living standards indicator, which is defined by the size of average life duration;

Y_W – reduced annual damage that is connected with decreasing population employ ability indicator.

The following expressions can be used for M_{COND}, Y_G, Y_Q and Y_W:

$$M_{COND} = \sum_{i=1}^{n} \frac{A_i}{m_i}, \tag{13}$$

where m_i – annual emission mass of the i-th admixture to the atmosphere (t/year); A_i – relative aggressiveness dimensionless parameter of the i-th admixture (cond.t/year).

$$Y_G = \left(\frac{\Delta N_{repr}^{cntr}}{N_{repr}^{cntr}} - \frac{\Delta N_{repr}^{ter}}{N_{repr}^{ter}} \right) \cdot N_{repr}^{ter} \cdot q_{repr}^{cntr}, \tag{14}$$

where ΔN_{repr}^{cntr} – reproductive age country population annual increase from 16 to 60;

N_{repr}^{cntr} – reproductive age country population at the target year beginning;
ΔN_{repr}^{ter} – reproductive age annual population growth;
N_{repr}^{ter} – reproductive age territory population at the target year beginning;
q_{repr}^{cntr} – specific gross national product, per head the country's reproductive population (tenge/person).

$$Y_Q = N_{repr}^{ter} \cdot q_{repr}^{cntr} \cdot \left(T_L^{cntr} - \frac{T_L^{ter}}{T_L^{cntr}} \right), \tag{15}$$

where T_L^{cntr} – average life duration in a country;
T_L^{ter} – average life duration in a territory.

$$Y_W = N_{repr}^{ter} \cdot q_{repr}^{cntr} \cdot \left(n_{repr}^{cntr} - n_{repr}^{ter} \right), \tag{16}$$

where n_{repr}^{cntr} – average annual number of man-days per reproductive age person in a country;

n_{repr}^{ter} – average annual number of man-days per reproductive age person in a territory.

The cost risk index is expressed as follows:

$$R_{VT} = Y_{norm} \cdot \left(\frac{M_{cond}^{AE}}{M_{cond}^{norm}}\right), \tag{17}$$

where M_{cond}^{AE} – possible amount of accidental emissions;

Y_{norm} and M_{cond}^{norm} – damage cost and the pollutant quantity in normative evaluation by the standard methodology.

The obtained general expression can allow to model and study the causes of changes in geoecological risks depending on all factors. This expression describes the environmental hazard value in relative terms; however the actually measured emissions concentrations data of harmful substances in the atmosphere are required for providing calculations. That is why calculation can be accomplished by analyzing the fact that damage has already occurred. Risk situations forecasting is impossible during the industrial plants performance using formula 8–17 because of the absence of C_i^{emis} values. The way out can be done by using theoretical formulas for calculating concentrations from an industrial plant's emissions stationary source. Then we can find an expression for the particular plant considering the air condition of a certain region [6], which can be substituted in place of C_i^{emis}.

The mathematical expression for the concentration from a point source with a constant power – Q (kg/s) for Ust-Kamenogorsk can be written as:

$$C(x,y,z,t) = \frac{f(A)\cdot Q}{2\pi\sigma_y\sigma_z U}\cdot\exp\left(-\frac{y^2}{2\sigma_y^2}\right)\cdot\left[\exp\left(-\frac{z-H^2}{2\sigma_z^2}\right)+\exp\left(-\frac{z+H^2}{2\sigma_z^2}\right)\right], \tag{18}$$

where Q – source power (kg/s);

σ_y and σ_z – dispersion parameters that depend on the atmosphere stability and the distance from the source "x" (m);

U – wind speed (m/s);

H – source height (m);

x, y, z – axial, transverse and vertical coordinates;

f(A) – impurity fraction in the mixing layer ("A"– mixing layer height).

The dispersion parameters σ_y yand σ_z were calculated by the formulas that were obtained by approximating the data for various atmospheric stability categories.

The calculating average annual concentrations problem solution in the residential zone of Ust-Kamenogorsk comes to the integration of all possible pollutants concentrations in a given point in space (x, y) and emission sources. Since it is assumed that within the M-rumba wind rose sector, the wind direction is fairly spread, which is typical for Ust-Kamenogorsk, the average annual concentration C (x, y) is calculated by the formula:

$$C(r,\theta) = \sum_{i=1}^{L} P_{Vi}\cdot\left\{\sum_{k=1}^{6}\left(P_k(U_i)\cdot\frac{M\cdot Q\cdot\gamma(x/U)}{2\cdot\sqrt{2}\cdot\pi^{3/2}\cdot r\cdot U_i\cdot\sigma_z}\cdot f(A,H,\sigma_z)\right)\right\}, \tag{19}$$

where Q – source power (kg/s);

P_{Vi} – wind realization probability at speed U_i (m/s) in the corresponding M-rumbling scheme sector;

$P_k(U_i)$ – realization probability of atmosphere stability certain class with the wind U_i (A-1, B-2, ..., F-6);

θ – wind direction in polar coordinates;

r – distance from the pollution source to the point (x, y);

σ_z – vertical dispersion characteristic;

$f(A, H, \sigma_z))$ – influence function of the pollution source height (H) and the mixing layer height (A);

$M/2\pi$ – sector angular fraction in the winds M-pattern;

$\gamma(x/U) = \gamma(t)$ – concentration change function along the plume axis due to photochemical reactions, dry and wet deposition, etc. in time.

Substituting the expression 19 in 8–17 as C_i^{emis}, and the expression 18 as C_i^{emisAE}, we will get an expression allowing to analyze the existing atmospheric pollution risk situation from an industrial plant, and to identify factors that have a greater contribution for certain plant features.

3 Practical Implementation

To study the risk impact on the environmental hazards value in the region, we developed the informational system for risk assessment using the above model, considering numerical calculations that are modeling the harmful substances concentration behavior. Which are trapped in the atmosphere as a result of industrial plant emissions taking into account the atmospheric stability and wind rose category in the region.

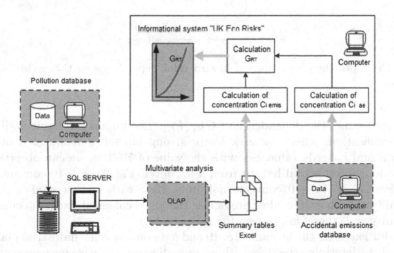

Fig. 1. General scheme of the informational system "UK Eco Risks".

The general scheme of the informational system "Ust-Kamenogorsk ecological risks" is presented in Fig. 1. The data from the atmospheric pollution database came to the SQL server, where the multidimensional cubes were built using OLAP technology and summary spreadsheets were created by slice and dice with the necessary information for us about the substances concentration upon which further calculation was continued. Information on emergency emissions came from the independent database in the separate informational system module. The quantitative analysis and the environmental hazard calculation were carried out in the region based on the incoming data.

4   Results

We analyzed the weight risk factor influence on the environmental hazard value in relative units based on the available data for Ust-Kamenogorsk [7].

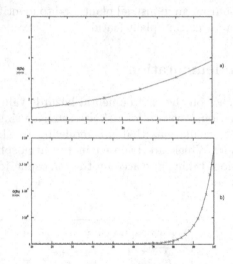

Fig. 2. Graphs of the environmental hazard value dependence on the environmental risk value.

Figure 2 shows the dependence of $G_{RT}(P_R)$. The important result is the P_R area identification, where the risk factor strong impact is manifested on the environmental hazards value. So, with the value of 70–100, we can observe the increase in environmental hazard from a low level (Fig. 2a) - 1–10 conventional units (Fig. 2b) to 106–107conventional units. Here with further studies on the probability of such events allow predicting the high ecological danger occurrence probability and, therefore, preventing this situation.

In addition using the formulas 18, 19 and data on a specific industrial plant, it is possible to determine the risks without providing quantitative measurements of pollutants in the studied region territory. Also the obtained information makes it possible to predict the ecological hazard occurrence that can lead to irreversible geoecological changes.

The most famous emergency situation that created unfavorable environmental conditions and entailed a significant increase in environmental danger appeared in the region in 1990, when, after the accident at the Ulba Metallurgical Plant, the beryllium cloud spread hundreds of kilometers and reached the territory of China (these emissions are the most significant for all history of Kazakhstan and world practice).

The pattern of the distribution of the emergency release plume is presented in Fig. 3 The dotted line indicates the area of beryllium detection, which is approximately 12 thousand square kilometers.

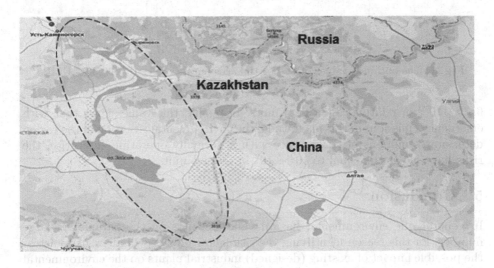

Fig. 3. Emission spread area.

In general, the magnitude of the environmental hazard depends on a large number of parameters; therefore, highlighting the influence of the risk factor will make it possible to realistically assess the possible impact of existing industrial enterprises, as well as being built or designed, on the appearance of environmental hazard in the region and the extent of possible violations associated with risk.

Having information about the size of the consequences, we solved the inverse problem of estimating the value of the proportionality coefficient (by the area of the gas cloud propagation, calculated the emission power), and carried out calculations of the environmental risk. The values obtained for the 1990 JSC UMP accidental release are shown in Fig. 4 by dashed lines.

Thus, the calculated value of environmental risk in relative conditional units is 972 units and shows that the values calculated on real data are in the zone of a large growth gradient of the indicator of environmental hazard and are

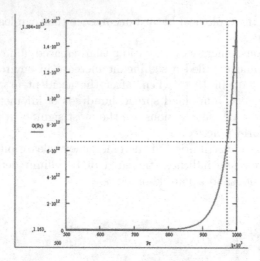

Fig. 4. The obtained values of P_R for the accidental release of 1990 JSC UMP.

$6,79 * 10^{12}$ in relative conditional units. In reality, this value shows that with emergency emissions of this magnitude, the environmental situation in the region deteriorates millions of times, since with a normally operating enterprise, the risks have indicators equal to $10^2 - 10^3$ in relative units.

5   Conclusion

In general, the environmental hazard value depends on the parameters large number, for this reason identifying the risk factor impact will allow us to assess the possible impact of existing (designed) industrial plants on the environmental hazards occurrence in the region, and to predict the possible violations extent associated with risk.

Conclusions and recommendations can be widely used in cities prone to the negative effects of industrial emissions. A new approach to estimating the risks of emissions from industrial enterprises and their impact on the magnitude of damages is proposed. The program complex and the information system make it possible to develop practical recommendations and evaluate the effectiveness of environmental protection measures in a new way.

References

1. Parizhskoe soglashenie po klimatu [Electronic resource]. http://ratel.kz
2. Zelenaia energetica Kazakhstana v 21 veke: mifi, realnost i perspectivi [Electronic resource]. http://gbpp.org/category/publication
3. Ibragimov, M.H.-G., Kutsenko, V.V., Rachkov, V.I.: Scientific foundations of the methodology of quantitative analysis of environmental hazards in the event of anthropogenic impact on the environment. Environmental Impact Assessment. Ecological Expertise 5, p. 23 (1999)

4. Kurilenko, E.A., Baklanov, A.E., Kvasov, A.I., Baklanova, O.E., Dmitrieva, T.S.: Development of an information system for studying the impact of industrial emissions on public health. In: Materials of the International Scientific and Methodological Conference 2 (2007)

5. Baklanov, A.E., Baklanova, O.E., Titov, D.N.: Influence of emissions of harmful substances in atmosphere on population health. In: International Forum on Strategic Technology, IFOST 2012 (2012)

6. Doudkin, M.V., Pichugin, S.Y., Fadeev, S.N.: Contact force calculation of the machine operational point. Life Sci. J. **10**(10s), 246–250 (2013)

7. Kim, A., Doudkin, M.V., Vavilov, A., Guryanov, G.: New vibroscreen with additional feed elements. Arch. Civil Mech. Eng. **17**(4), 786–794 (2017)

Using an Ontological Model for Transfer Knowledge Between Universities

M. Bazarova(✉), G. Zhomartkyzy, and S. Kumargazhanova

D. Serikbayev East Kazakhstan State Technical University,
Serikbaev Street, 19, 070010 Ust-Kamenogorsk, Republic of Kazakhstan
madina9959843@gmail.com, zhomartkyzyg@gmail.com, skumargazhanova@gmail.com

Abstract. The article deals with construction of the ontological model to provide knowledge transfer between universities. It discusses a need to consider requirements of the professional standard for a specialist and labor market demand in process of educational programs design. Ontological and modular-competence approaches were used to simulate the mapping of competences. An ontological model of the professional competence standard was developed. The information model distributed knowledge database in the university consists of the ontology educational programs, the ontology of professional competencies demanded in the regional labor market and the ontology of scientific knowledge in the university. Classes taxonomy of the developed ontological model of professional competences is presented. The method of analyzing hierarchies was used to assess the conformity of learning outcomes with the functional requirements of labor market.

Keywords: Educational program · Competence · Ontology ·
Professional standard · Modular-competence approach ·
The method of analyzing hierarchies

1 Introduction

A notion of "competence" supplements triad "Knowledge-experience - skills". The triad is not sufficient to describe the educational process. It is necessary to consider professional competencies for providing labor market with highly qualified specialists. "Professional competencies" mean to master knowledge, skills and abilities in professional field The concept of academic competence of a graduate is same as the definition given in the European project Tuning. Academic competence is the ability to know and understand the theoretical knowledge in academic field; the ability to apply knowledge to specific situations; the values as an integral part of the way of perception and life in a social context [1–4]. Competence is a set of skills specific to each individual including qualification. It means that qualification is not enough to be a result of education. The competence approach is a priority orientation goals to the vectors of education: learning, self-determination (self-determination), self-actualization, socialization

© Springer Nature Switzerland AG 2019
Y. Shokin and Z. Shaimardanov (Eds.): CITech 2018, CCIS 998, pp. 34–43, 2019.
https://doi.org/10.1007/978-3-030-12203-4_4

and the development of individuality [5]. The modular-competence approach allows to move from knowledge to application and organization of knowledge it also increases the flexibility of the education system to expand the possibility of performing work. The Republic of Kazakhstan as the leading world states supports the concept of a market-oriented innovative university based on a triangle of knowledge (education-science-innovation). The concept is aimed to large-scale investment in human resources, development of professional skills and research, supporting the modernization of the education system in order to meet their needs of a global knowledge-based economy [6]. To implement innovative activities there is a need in a system of its organization which was named as Knowledge Transfer System [7]. The Bologna process unambiguously determines the need to adjust the system of relationship between universities and enterprises considering transfer of knowledge as the key component of university development, which provides commercialization of research and market - oriented educational programs. Knowledge transfer is organizational systems and processes through which knowledge, including technology, experience and skills is transferred from one side to the other which leads to innovation in the economy and social sphere [8]. The article considers application of the modular-competence approach to educational programs, taking into account the requirements of professional standards. The modular-competence approach with use of ontology to educational programs is reflected enough in works [9–12].

2   The Ontological Model

Agreement between educational programs and professional standards ensures high-quality training of specialists. It means the correspondence of academic and professional competencies. The more this correspondence is the more qualitative will be the training of specialists. Correspondence to European framework of ICT competencies and agreement of our educational programs to global standards are essential for our specialists to compete in global labor market. Comparison of the competencies of educational program, professional standards and the European Framework of ICT Competencies (the European e-Competence Framework - e-CF) form the general requirements for vocational training required to carry out specific work activities. Competencies of the educational standard are the results of training (competence) from the modular educational program of the specialty. To achieve the planned result of learning it is necessary to study certain modules with subjects. Competences of professional standards correspond to knowledge, experience and skills necessary to work in professional phere. Competencies from the e-CF qualifications framework are reference competencies that include knowledge and skills. The ontological approach allows us to combine these three types of competences and transfer professional competences to the content of subjects. It will ensure the demand for graduates of higher educational institutions in the labor market without additional training or retraining. An important function of the ontological model is the integration of heterogeneous data and knowledge of various fields of knowledge. The ontological model differs from the others written using known means and graphical languages by them being not just individual models, but fragments of a general ontology [13,14]. The information model of the university's distributed knowledge base includes the ontology of the university's educational programs, the ontology of the professional competencies demanded on the regional labor market and the ontology of the European ICT Competence Framework (e-CF) [15]. The model of knowledge of the university is a set of ontologies, has the following form 1:

$$O_U = <O_{EP}, O_{PS}, O_{e-CF}>$$ (1)

where

- O_{EP} - ontology of educational programs of the university;
- O_{PS} - ontology of professional standards;
- O_{e-CF} - ontology of the European ICT Competence Framework.

The ontological approach allows us to combine these three types of competences and transfer professional competences to the content of subjects which will ensure the demand for graduates of higher educational institutions in the labor market without additional training or retraining. Using the ontological approach, it is necessary to develop a knowledge base of modular educational programs of the university, professional competencies and e-CF. Using the ontology of educational programs EKSTU allows to automate the stages of expertise associated with verifying the composition and structure of modular educational programs, checking the consistency of the prerequisites for modules. The ontological model of educational programs, created earlier in the framework of the Grant Financing project, has been improved with the help of new concepts, properties and relationships. To build a complete information model of university knowledge, it is necessary to build ontologies of professional competencies and e-CF. The article deals with the implementation of the modular-competence approach in the field of information and communication technologies. In the Republic of Kazakhstan, the sectoral framework of qualifications and professional standards in the field of information technology (IT) are under development. On the basis of them universities would develop educational programs for training IT professionals. Holding company "Zerde" develops 10 professional standards, which will create the requirements for the professional qualification of IT professionals. Professional standards reflect professions, for each profession, skill levels and labor functions are defined. Based on the structure of the Professional Standard of the Republic of Kazakhstan, the main classes and relations of ontology are defined, the ontology of professional standards is constructed Fig. 1 Ontology, together with many individual instances of classes, constitute a knowledge base. Figure 2 shows a fragment of the constructed ontology.

3   Assessment of the Compliance of Training Results with the Qualification Requirements of Labor Functions

The method of analyzing hierarchies (MAI) involves decomposing the problem into ever simpler parts and handling expert judgments. As a result, the relative

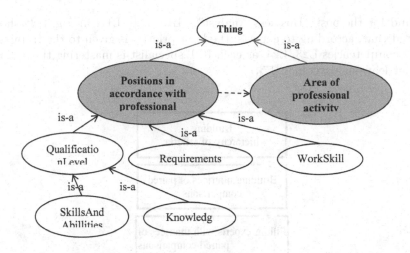

Fig. 1. The main classes of the ontology of the professional standard

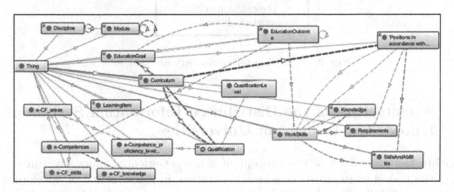

Fig. 2. Fragment of the constructed ontology

significance of the investigated variants is determined for all criteria in the hierarchy [16]. Relative significance is expressed numerically in the form of priority vectors. To assess the compliance of training results with the qualification requirements of labor functions for a particular post, it is necessary to perform several steps, shown in Fig. 3. Matrices of paired comparisons are constructed and filled by experts. The calculation and hierarchical synthesis have been made. The analysis of the results of the resultant vector shows which options (learning outcomes) are prioritized for this position. As follows from the diagram in Fig. 4, the highest priority for the position of "Information Security Specialist" has, according to experts, the result of training (competence), according to L12 special competence. The highest priority for the post "Network Administrator" has the result of training (competence), formulated in L14, L11, L13 special competencies. The highest priority for the position of "Software Developer" (Fig. 4) has, according to experts, the result of training special competencies L11, L13,

L14, and for the post "Business Analyst" - L19, L20, L17. in Fig. 4. It should be noted that, according to experts, the least priority is given to the training of special competencies L23, L24, as each ICT specialist is mastering them, based on other learning outcomes [17].

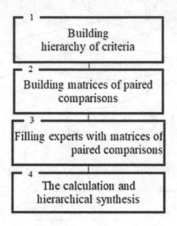

Fig. 3. Algorithm for conformity assessment

4   Architecture of the Distributed Information System of Knowledge Transfer of Universities

Within the framework of the concept of a market-oriented innovative university, the task is to improve the education system, in the direction of developing professional skills. To solve this problem, it is necessary to build a distributed information system for the transfer of knowledge of higher education institutions Fig. 5. Participants of knowledge transfer of universities are students, university entrants, teachers, researchers, employers, management, the authorized education and science bodies and the administrator. The general information model of the university's distributed knowledge base includes the ontology of the university's educational programs, the ontology of professional competencies demanded inthe regional labor market, the ontology of the European ICT Competence Framework (e-CF) and the ontology of scientific knowledge. The ontology of scientific knowledge was developed earlier in the framework of the Grant Financing Project. Distributed systems are client-server systems. The models of these systems and advantages are considered in the article [11]. The three-link model is favorable becausethe interface with the user is completely independent of the data processing component. In the three-tier model, the user interface component and data management component (and databases including) are clearly highlighted.. Between them there is middleware which performs the functions of managing transactions and communications, transporting requests, managing names and many others. Middleware is the main component of distributed

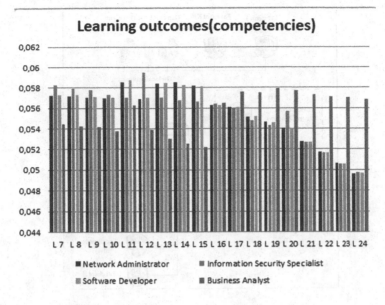

Fig. 4. Learning outcome

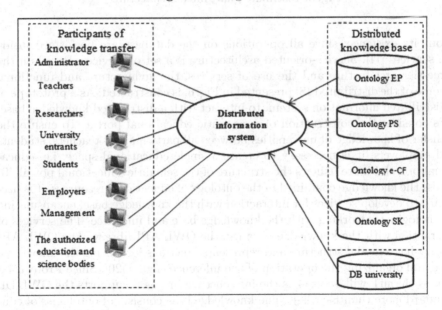

Fig. 5. Distributed integrated circuit

systems. The client explicitly requests one of the services provided by the application component. The client sends the request to the information bus, without knowing anything about the location of the service. For the Client, the database is hidden by a layer of services. Moreover, he generally does not know anything

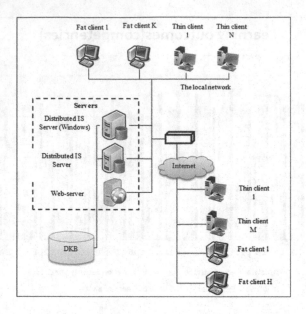

Fig. 6. Distributed integrated architecture

about its existence, since all operations on the database are performed inside the services [11]. Service-oriented architecture is a set of services. Based on the three-tier architecture and the use of services, the architecture and functional scheme of the distributed IS, presented in Fig. 6, is constructed. As a prototype of a distributed information system, to interact with a distributed knowledge base, let's consider the construction of a semantic educational portal. To ensure the transfer of knowledge of universities between all participants - teachers, students and employers, it is necessary to create a single educational space, i.e. educational portal. Figure 7 shows the structure of the semantic educational portal. To access the knowledge contained in the ontology of educational resources, it is necessary to develop services for interaction with the knowledge base, encapsulating technologies for working with the knowledge base and inference. The services of interaction with the knowledge base use the OWL API library. The OWL API library has a modal structure and represents functions for editing the knowledge base and organizing the operation of the inference engine [20]. Since Protégé 4.3 is used HermiT will be used as the inference engine, as it supports the OWL DL standard more than Fact ++. The knowledge base consists of ontologies of educational programs, professional standards and the European Framework for ICT Competencies (e-cf) Fig. 7 The main components are a distributed knowledge base management system (SURB), a web server and an application server (middleware). To access and manage data, a link is used for "thin" or "thick" clients. The distributed knowledge base assumes the storage and execution of knowledge management functions in several nodes and the transfer between these nodes during the execution of requests. Partitioning of data in a distributed

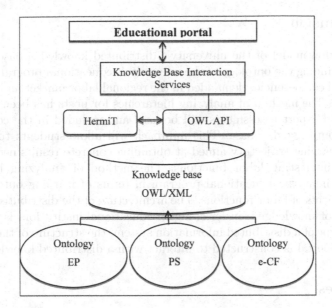

Fig. 7. The main classes of the ontology of the professional standard

database can be achieved by storing various tables on different computers or even storing different parts and fragments of the same table on different computers. For a user (or application program), it does not matter how knowledge is shared between computers. To work with a distributed knowledge base, if it is really distributed, should be the same as with a centralized database, i.e., the location of the database should be transparent [18]. To access the local network for convenience, a Windows server can be used. The knowledge base is located on another server, for greater security, it is recommended to use a Unix server [19]. Two servers are used to work with the knowledge base. The user has the opportunity to work with distributed IP, not only in the local network, but also via the Internet, using the Web server. For access through the local network, LAN (LAN) support is used. For Internet access, a web browser and network protocols are used. The competency-based approach will help to overcome the discrepancy between the requirements for the quality of education between the state, society and the employer. That in its turn will allow the university to respond flexibly to changing conditions of the external environment and develop educational programs, improving their quality and relevance in the market of educational services. It is essential to create a single educational space, i.e. educational portal to ensure the transfer of knowledge of universities between all participants - teachers, students and employers.

5   Conclusion

An information model of the university's distributed knowledge base has been built up, including the ontology of the university's educational programs and the occupational competencies demanded in the regional labor market and the ontology of e-CF. The method of analyzing hierarchies for posts has been approved. The obtained expert assessments will be used and included in the competence knowledge ontology database. This approach will allow students to build an individual learning trajectory aimed at obtaining concrete results necessary for performing interesting labor functions. The method of analyzing hierarchies allows examining the educational program in terms of learning outcomes and the requirements of labor functions. The architecture of the distributed information system of knowledge transfer of higher educational institutions is presented. As a prototype of a distributed information system, the structure of the semantic educational portal is constructed to interact with a distributed knowledge base.

References

1. Uvalieva, I., Smailova, S., Tarifa, F.: The development of an informational analysis software for the monitoring of distributed objects within socioeconomic systems. Actual Probl. Econ. **3**(165), 482–491 (2015)
2. Uvalieva, I., Chettykbayev, R., Utegenova, A., Toibayeva, S.: Mathematical basis and information system software for educational institutions ranking. In: Proceedings of the International Conference Application of Information and Communication Technologies-AICT 2015, Rostov-on-Don, Russia, pp. 487–490 (2015)
3. Zhomartkyzy, G., Balova, T.: The development of information models and methods of university scientific knowledge management. In: Hammoudi, S., Maciaszek, L., Teniente, E., Camp, O., Cordeiro, J. (eds.) ICEIS 2015. LNBIP, vol. 241, pp. 429–451. Springer, Cham (2015). https://doi.org/10.1007/978-3-319-29133-8_21
4. Zhomartkyzy, G., Balova, T.: Monitoring the development of university scientific schools in university knowledge management. In: Proceedings of the 17th International Conference on Enterprise Information Systems - ICEIS 2015, Barcelona, Spain, 27–30 April 2015, vol. 2, pp. 222–230 (2015)
5. Zeer, E.F.: Kompetentnostnyiy podhod k obrazovaniyu. Obrazovanie i nauka (2005)
6. Gorbunova, E.M., Larionova, M.V.: Evolyutsiya problematiki obrazovaniya v kontekste prioritetov i obyazatelstv Gruppyi vosmi, Universitetskaya kniga, Logos, pp. 77–102 (2007)
7. Hagen, S.: From tech transfer to knowledge exchange. In: Wenner-Gren International Series. The University in the Market, vol. 84, pp. 103–117. Portland Press, London (2008)
8. Grudzinskiy, A.O., Bednyiy, A.B.: Transfer znaniy - funktsiya innovatsionnogo universiteta. Vyisshee obrazovanie v Rossii, pp. 66–71 (2009)
9. Andreeva, N.M.: Model informatsionno-kommunikatsionnoy kompetentnosti studentov biologicheskih i ekonomicheskih spetsialnostey. Vestnik Krasnoyarskogo gosudarstvennogo pedagogicheskogo universiteta im. VP Astafeva, no. 1 (2015)
10. Bolotov, V.A., Serikov, V.V.: Kompetentnostnaya model: ot idei k obrazovatelnoy programme. Pedagogika **10**, 8–14 (2003)

11. Verhoturova, Y.S.: Ontologiya kak model predstavleniya znaniy. Vestnik Buryatskogo gosudarstvennogo universiteta **15** (2012)
12. Gafurova, A.G., Piyavskiy, S.A.: Planirovanie i organizatsiya uchebnogo protsessa v vuze pri kompetentnostnoy podgotovke studentov. Transportnoe delo Rossii **3** (2013)
13. Mukhacheva, N.N., Popov, D.V.: Ontological models and methods for managing information and intellectual resources of an organization. Bulletin of Ufa State Aviation Technical University 14, vol. 1, no. 36, pp. 123–135 (2010)
14. Berners-Lee, T., Hendler, J., Lassila, O.: The semantic web. Sci. Am. **284**(5), 34–43 (2001)
15. Doudkin, I.M.V., Pichugin, S.Y., Fadeev, S.N.: Contact force calculation of the machine operational point. Life Sci. J. **10**, 246–250 (2013)
16. Saaty, T.: Decision Making. Hierarchy analysis method, Radio and communication (1993)
17. Bazarova, M.Z., Zhomartkyzy, G., Wojcik, W., Krak, Y.V.: Construction of individual trajectories of training specialists in the field of information and communication technologies. J. Autom. Inf. Sci. **49**(10), 36–46 (2017)
18. Kudryaev, V.A., Korneev, I.K., Ksandopulo, G.N.: Organization of work with documents, pp. 22–25. INFRA-M, Moscow (1999)
19. Karpova, I.P.: Research and development of knowledge control subsystems in distributed automated learning systems. Thesis for the title of candidate. Tech. Sciences specialty, pp. 22–25. INFRA-M, Moscow (1999)
20. Doudkin, M.V., Vavilov, A.V., Pichugin, S.Y., Fadeev, S.N.: Calculation of the interaction of working body of road machine with the surface. Life Sci. J. **133**, 832–837 (2013)

Management of Oil Transportation by Main Pipelines

T. Bekibayev[1], U. Zhapbasbayev[1], G. Ramazanova[1(✉)], E. Makhmotov[2], and B. Sayakhov[2]

[1] Kazakh National Research Technical University after K.I. Satbayev,
Almaty, Kazakhstan
gaukhar.ri@gmail.com
[2] JSC "KazTransOil", Astana, Kazakhstan

Abstract. Integration of SmartTran software with a SCADA system allows real-time simulation, monitoring and optimization of oil pumping through main oil pipelines of the Republic of Kazakhstan. The results of the digital technology development to control oil transportation along the main oil pipelines of the Republic of Kazakhstan using the SmartTran software are presented. The calculations results of pumping modes prove the economic efficiency of digital technology implementation.

Keywords: Digital technology · Integration · Main oil pipeline ·
Transportation of oil mixtures · Monitoring · Modeling · Optimization

1 Introduction

The management of oil transportation via the main oil pipelines is carried out using the SmartTran software in conjunction with a SCADA system, which provides monitoring, management and optimization of technological modes of oil pumping [1,2]. This technology monitors the operation of pumping units, heating furnaces, oil pumping modes in real time and allows process automation during pipeline transportation. As a result, operational reliability and efficiency of the main oil pipelines is increased. In this paper, we present the results of the digital technology development for oil transportation through the main oil pipeline sections by integrating the SmartTran software and the SCADA system of JSC "KazTransOil" (hereinafter KTO).

2 SCADA System of the KTO

The main oil pipelines (MP) of the KTO are characterized by the following features:

- the main technological objects of MP have a high single capacity;
- MP objects are classified as dangerous;

Y. Shokin and Z. Shaimardanov (Eds.): CITech 2018, CCIS 998, pp. 44–53, 2019.
https://doi.org/10.1007/978-3-030-12203-4_5

- MP objects are spaced at long distances;
- climatic and other external conditions are severe, unfavorable.

The principle of operation of Supervisory Control and Data Acquisition (SCADA system) is in providing industrial and information security; effectiveness; standardization and information compatibility.

The SCADA system of the KTO is one of the largest in the world with technological equipment including 80 main pumping units, 40 booster pumps, 62 heating furnaces, 86 tanks, 626 auxiliary system, 1731 pipe gate valves, 2468 temperature sensors, 1649 pressure sensors, 995 level sensors, 239 consumption sensors, 323 vibration sensors, 213 gas content sensors, 721 current and voltage sensors.

The main function of the SCADA system:

- visualization and information input;
- data control;
- management of process equipment;
- registration and storage of events and accidents;
- storing the history of technological parameter values;
- reports formation;
- communication with technological networks (via OPC technology);
- inter-level transmission of technological information.

The SCADA system receives data via the fiber-optic system from sensors that measure pressure and temperature of the oil flow, the soil temperature at the points of measurement in the linear section of the pipeline and equipment (pump units, gate valves, etc.). The values from pressure, temperature and flow sensors helps to monitor the flow of oil in the pipeline and the operation of process equipment.

3   SmartTran Software

The SmartTran software is a joint development of the authors, and it is designed for forecasting, modeling and optimization of energy-saving modes of oil mixtures transport at sections of the main oil pipelines. Also it gives an opportunity to design new pipelines with the identification of sections, pumping equipment and heating furnaces. Modeling and optimizing the energy saving mode of "hot" oil transportation distinguishes the SmartTran software from other products.

The functionality of the SmartTran software consists of the following tasks:

1. Determination of energy-saving operation modes of pumping units with detachable rotors and variable frequency drive;
2. Determination of the optimum oil mixtures temperature at pipeline sections under energy-saving operating conditions for heating furnaces;
3. Determination of energy saving modes of "hot" oil transportation in main oil pipelines;

4. Determination of energy saving modes of oil transportation using chemical additives (depressor and anti-turbulent additives);
5. Determination of the economic efficiency of technological modes of oil transportation taking into account the difference in tariffs for fuel and electricity.

SmartTran performance capabilities:

1. Heat-hydraulic calculations of stationary modes of high viscous and high pour point oil transportation for the safe operation of the main pipelines (taking into account associated pumping and pumping out, loops, pipe defects, pressure regulator and input of additives);
2. Heat-hydraulic calculations of non-stationary pumping modes after short-term stops for the safe operation of main oil pipelines;
3. Heat-hydraulic calculations of serial transfer of different varieties of high viscous and high pour point oils mixtures along main oil pipelines;
4. Heat-hydraulic calculations of energy saving modes of main pumping units with detachable rotors and variable frequency drive for stationary operation;
5. Determination of the optimum temperature of heating oil mixtures and energy-saving modes of heating units for stationary operation;
6. Selection of pumping equipment of pumping stations (PS) with detachable rotors and variable frequency drive for forecasting the maximum capacity of the pipeline with permissible operating modes of pumping equipment;
7. Adaptation of real characteristics of pumping equipment of the PS according to the SCADA system data;
8. Designing of new sites, the addition of pumping equipment at PS and input of rheological properties of oil mixtures.

Figure 1 shows the SmartTran software window during calculation of the main pumping station (MPS) Uzen - PS named after T. Kasymov section.

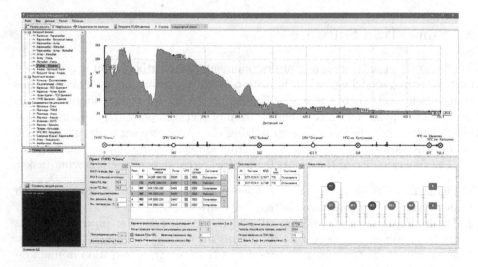

Fig. 1. A SmartTran software window during calculation of the selected section

4 Integration of SmartTran with SCADA

Figure 2 shows structurally how the SmartTran integrates with the SCADA system.

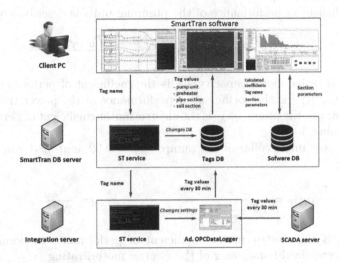

Fig. 2. Integration diagram of the SCADA system with SmartTran

As a result of integration the following initial data were obtained:

1. Pressure-volumetric characteristics of pumps, depending on the service life for determining the power consumption and efficiency;
2. Pressure and temperature values at the inlet/outlet of pumping units and oil pumping stations;
3. Oil parameters in the linear sections of the main oil pipeline (flow rate and pressure, oil temperature);
4. Hydraulic and thermal characteristics of pipes within sections (waxing of inner part, stagnant zones, soil parameters);
5. Parameters of the heating furnaces (oil and gas consumption, oil temperature at the inlet and outlet of the furnaces).

The SmartTran software uses data obtained from the SCADA system to determine the energy saving modes of oil mixture transportation at sections of the main oil pipelines.

5 Determination of Pumping Units Power Consumption

Figure 3 shows the head-capacity characteristic of the pumping units, taking into account its operational life, obtained by the SmartTran adaptation module. Actual data of volumetric flow and pressure drop (Q, $Pout - P_{in}$) obtained by the

SCADA system are used to determine the power consumption and the coefficient of performance of the pumping unit. The points indicate an actual data of the pumps working. The transition from blue to red color displays an increase in the points concentration. The adaptation module constructs a regression curve (black line on the plot).

The coefficient of performance of the pumping units is described by [3]:

$$\eta_{PE} = \eta_P \eta_{MT} \eta_{EM} = \frac{(Pout - P_{in})Q}{N_{act}} \qquad (1)$$

where N_{act} is active motor capacity, η_P is the coefficient of performance of the pumping unit, η_{MT} is the coefficient of performance of the power transmission from the motor to the pump, η_{EM} is the electric motor coefficient of performance, Q is the volume flow.

For the mechanical collar of the pump $\eta_{MT} = 0.99$, and η_{EM} can be calculated by [1]:

$$\eta_{EM} = \frac{1}{1 + \frac{1-\eta_{nom}}{2\eta_{nom}k_{load}}(1 + k_{load}^2)}, \quad k_{load} = \frac{N_{act}}{N_{nom}} \qquad (2)$$

where η_{nom} is the electric motor coefficient of performance at nominal load, N_{nom} is the standard horsepower of the electric motor, rating.

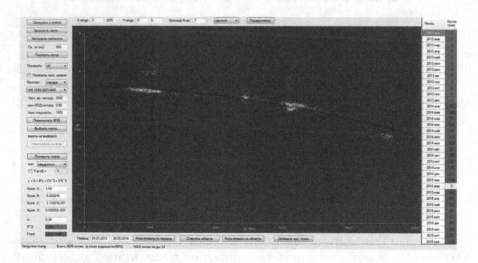

Fig. 3. Pressure drop vs volumetric flow rate relationship dP(Q)

Based on the SCADA system's data the coefficient of performance of the pump is calculated as:

$$\eta_P = \frac{(P_{out} - P_{in})Q}{N_{act}\eta_{MT}} \left(1 + \frac{1 - \eta_{nom}}{2\eta_{nom}k_{load}}(1 + k_{load}^2) \right) \qquad (3)$$

and energy consumed by pumping units:

$$N_P = \frac{\rho g H Q}{\eta_P \eta_{EM} \eta_{MT}} \tag{4}$$

where H is the pump head, g is the gravity acceleration.

Power consumption data are used to optimize energy saving modes of pumping units.

6 Determination of the Energy Consumed by the Heating Furnace

Calculation of the fuel consumption of any type is carried out in units of reference fuel. As a reference fuel is used 1 kg of fuel with a lowest calorific value $Q^p_{r.f.} = 7000\,\text{kcal/kg}$ (29.3 MJ/kg). Specific consumption of gaseous fuel for heating 1 ton of oil by 1 °C in the i-th heater is found from expression [4]:

$$b^i_f = \frac{7000}{Q^H_p} \cdot b^i_{r.f.}, \ b^i_{r.f.} = \frac{142.86 \cdot 10^{-3}}{\eta^\theta_i} \cdot c_p \tag{5}$$

where Q^H_p is the calorie power of gas fuel, $b^i_{r.f.}$ is the specific consumption of reference fuel, c_p is the heating capacity of oil, η^θ_i is the coefficient of performance of the i-th heater.

Knowing the specific fuel consumption, it is possible to determine the amount of gas G^{gas}_i (10^3 Nm^3) for heating specific amount of oil G^{oil}_i (10^3 tons) on the i-th heater [4]:

$$G^{gas}_i = b^i_{r.f.} \cdot G^{oil}_i \cdot (T^{ex}_i - T^{en}_i) \tag{6}$$

where T^{ex}_i, T^{en}_i are oil temperature at inlet and outlet of the i-th heater.

In the case when several heating furnaces operate at a heating station the total amount of fuel is found from the following expression:

$$G^{gas} = \sum_i^n b^i_f \cdot G^{oil}_i \cdot (T^{ex}_i - T^{en}_i) = \frac{7000}{Q^H_i} \sum_i^n \frac{0.14286}{\eta^\theta_i} c_{pi} \cdot (T^{ex}_i - T^{en}_i) \cdot G^{oil}_i \tag{7}$$

The SCADA system data determine the temperature of oil at the i-th heating furnace inlet and outlet, at the heating station inlet and outlet; pressure at the heating station inlet and outlet, and the oil flow through the i-th furnace.

Based on SCADA system data SmartTran carry optimization of fuel consumption for heating oil at the heating station during "hot" transportation.

7 Determination of the Hydraulic Resistance of the Pipeline

Hydraulic resistance is the most important characteristic of the pipeline, and the accuracy of its determination affects the economic efficiency of oil transportation.

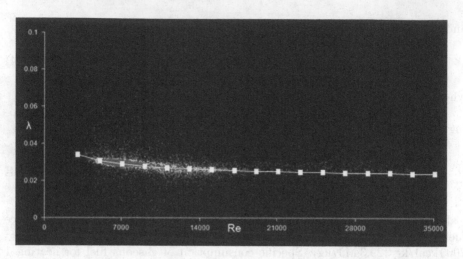

Fig. 4. Dependence of the coefficient of hydraulic resistance on Reynolds number $\lambda(Re)$

Hydraulics of the oil pipeline depends on many factors (viscosity, roughness, flow velocity) when pumping high pour point (paraffinic) and high viscous oil and is based on the Darcy-Weisbach formula [1–3]. The oil temperature varies along the length of the pipeline due to heat transfer with the soil, and as a result the oil viscosity changes. The pipe roughness can vary for various reasons, including wax deposition on the walls, and requires constant adaptation of the hydraulic resistance.

The coefficient of hydraulic resistance is determined by solving the system of motion and continuity equations [5]:

$$\rho_0 \frac{\partial w}{\partial t} + \frac{\partial p}{\partial x} = -\lambda \frac{\rho_0 w^2}{2D_1} + \rho_0 g \sin \alpha(x) \tag{8}$$

$$\frac{\partial p}{\partial t} + \rho_0 c^2 \frac{\partial w}{\partial x} = 0 \tag{9}$$

where p, ρ_0, w are pressure, density, velocity of oil, g is gravity acceleration, λ is hydraulic resistance coefficient, D_1 is internal diameter of pipeline, α is angle of inclination of the pipeline axis to the horizontal, c is the speed of wave propagation in the pipeline ($c \approx 1000\,\mathrm{m/s}$).

The coefficient λ is expressed by the modified Altshul formula [1]:

$$\lambda(Re) = a \left(\frac{68}{Re} + e \right)^b + d \tag{10}$$

In the formula (10) the coefficients (a, b, e, d) are considered unknown and are found by comparing calculated and experimental data.

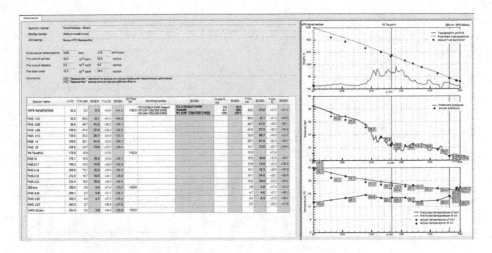

Fig. 5. Comparison of data from the SmartTran software and the SCADA system for the Karazhanbas-Aktau section

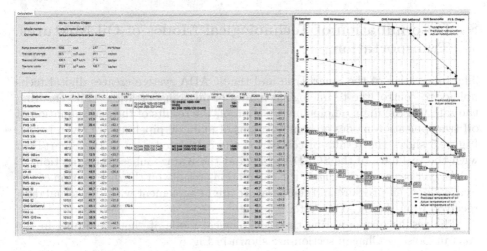

Fig. 6. Comparison of data from the SmartTran software and the SCADA system for the Atyrau-Bolshoy Chagan section

As indicated above, the SmartTran adaptation module determines the coefficient of hydraulic resistance by comparing the calculated and experimental data of the SCADA system. After that dependence of the coefficient of hydraulic resistance on Reynolds number $\lambda(Re)$ at the section of the main oil pipeline is identified (Fig. 4). In the Fig. 4 square dots indicate the experimental data of the SCADA system, the white line is a regression curve.

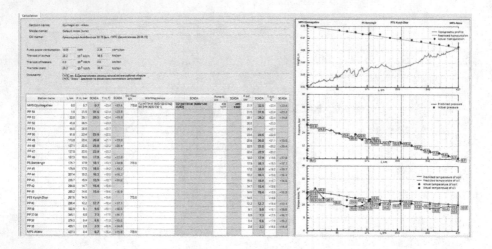

Fig. 7. Comparison of Smart Tran and the SCADA system data at the Djumagaliev-Atasu section

8   Determination of Technological Modes of Oil Mixtures Transportation

The integration of the SmartTran and the SCADA system creates digital technology to control the technological modes of oil transportation while ensuring the safety of the main oil pipeline. As a proof, the results of the determination of oil transportation modes on some sections of main oil pipelines are given below. Figure 5 shows the results of comparing the calculated data of the SmartTran with the real data of the SCADA system at the Karazhanbas-Aktau section.

As can be seen from Fig. 5, the calculations results on the distribution of the hydro-slope, oil pressure and temperature are in agreement with the actual data received from the SCADA system.

The results of monitoring the technological mode of "hot" pumping in the Atyrau-Bolshoy Chagan section are shown in Fig. 6.

It is necessary to note the agreement of the data calculated at the Smart Tran and the field data received from the SCADA system regarding power capacity of pumps at the PS named after T. Kasymov and Inder (Fig. 6).

Similar data were obtained at the Djumagaliev - Atasu (Fig. 7) and Djuma-galiyev - Chulak Kurgan (Fig. 8) sections.

Thus, the developed digital technology allows the simulation of oil pipeline operation and manages the technological modes of oil mixtures transportation through the integration of the SmartTran software and the SCADA system.

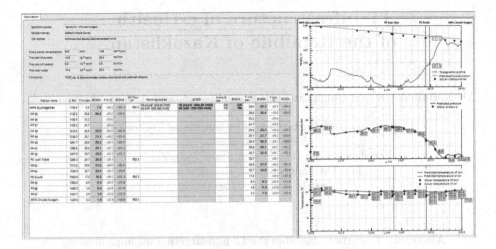

Fig. 8. Comparison of Smart Tran and the SCADA system data at the Djumagaliyev-Chulak Kurgan section

9   Conclusions

1. As a result of integration of the SCADA system and the SmartTran a digital technology to control and manage the technological modes of oil transportation through the main oil pipelines of the Republic of Kazakhstan was created.
2. The results of modeling using the SmartTran software are in accordance with the field data of the SCADA system. Digital technology based on the Smart-Tran software and the SCADA system is an effective tool for efficient operation of the main oil pipelines of the Republic of Kazakhstan.

We would like to thank the Ministry of Education and Science of the Republic of Kazakhstan for financial support through grant AP05130503 for 2018–2020.

References

1. Beisembetov, I.K., Bekibaev, T.T., Zhapbasbayev, U.K., et al.: Management of energy-saving modes of oil mixtures transportation by the main oil pipelines. KBTU, Almaty (2016)
2. Zhapbasbayev, U.K., Makhmotov, E.S., Ramazanova, G.I., et al.: Calculation of the optimal pumping temperature for oil transportation. Sci. Technol. Pipeline Transp. Oil Oil Prod. **20**, 61–66 (2015)
3. Tugunov, P.I., Novoselov, V.F., Korshak, A.A., Shammazov, A.M.: Typical calculations for the design and operation of gas and oil pipelines. Design-Polygraphservice, Moscow (2002)
4. Kuznetsov, A.A., Kagermanov, S.M., Sudakov, E.N.: Calculations of processes and apparatuses of the oil refining industry. Chemistry, Leningrad (1974)
5. Lurie, M.V.: Modeling of oil product and gas pipeline transportation. Wiley-VCH Verlag GmbH & Co. KGaA, Weinheim (2008)

IT Infrastructure of e-Health
of the Republic of Kazakhstan

M. Kalimoldayev[1], S. Belginova[2(✉)], I. Uvaliyeva[2], and A. Ismukhamedova[2]

[1] Institute of Information and Computational Technologies,
Almaty, Kazakhstan
[2] D. Serikbayev East Kazakhstan State Technical University,
Ust-Kamenogorsk, Kazakhstan
saule.belginova@mail.ru

Abstract. The article describes the IT infrastructure for implementing e-health based on the development of standards that determine the feasibility of realization a service-oriented architecture. This IT infrastructure provides full interoperability between information systems involved in supporting healthcare processes. Particular attention was paid to the introduction, use and features of a unified health information system. Also features of architectural models and management characteristic for e-health of Kazakhstan are described. The functional and architecture of e-health information systems are considered; the proposed conceptual functional architecture for health care; architecture of software and hardware for e-health.

Keywords: E-healthcare · Medical information system ·
Information system architecture · Infrastructure ·
Electronic health passport

1 Introduction

In the modern world, one of the main factors determining the dynamics of social and economic development of any state, including health care, is the informatization of society and the active introduction of information and communication technologies in all sectors of human life.

The main goal of informatization of healthcare in general can be formulated as follows: the creation of new information technologies at all levels of health management and the introduction of new medical computer technologies that improve the quality of medical and preventive care and contribute to the realization of the basic function of protecting public health - to increase the duration of active life [1].

In order for health services to promote the integration of medicine and public health, they must have three main characteristics: a population-based and geographical focus in the context of a decentralized health system; an attempt to

Y. Shokin and Z. Shaimardanov (Eds.): CITech 2018, CCIS 998, pp. 54–63, 2019.
https://doi.org/10.1007/978-3-030-12203-4_6

develop organizational models to support coordination and integration processes; and the proper use of the health information system.

As we know, the medical industry produces and accumulates a very large volume of medical and statistical data. All these data are to some extent used by doctors, medical workers, and also by the governing bodies in order to properly organize and improve the quality of medical care and raise the overall standard of living of the population.

To be relevant, a health information system must fit into the organization of the health system for which it generates information. Based on clearly defined management functions, it is relatively easy to identify the information needed to make appropriate decisions at each level of management.

The next question is how to obtain this information in the most efficient and effective way. To answer this question, it is important to understand the structure of the health information system.

Health information system, like any system, has an organized set of interrelated components that can be grouped into two objects: information process and management structure. Through the information process, raw data (input data) are converted into information in a "convenient" form for management decision-making (output data). The information process can be broken down into the following components: (I) data collection, (2) data transmission, (3) data processing, (4) data analysis and (5) reporting for use in patient and health care solutions for service management.

Not only health of the population depends on the qualitative use of medical information, but also the level of development of the country as a whole. Thus, the use of constantly growing large volumes of medical data in solving diagnostic, therapeutic, statistical and management tasks for the development of e-health in Kazakhstan is topical.

The prerequisites for the development of e-health in the Republic of Kazakhstan are:

(1) The State Program of Health Care Reform and Development for 2005–2010.
(2) The Code of the Republic of Kazakhstan "On the health of the people and the health care system".
(3) The State Program for the Development of Health Care of the Republic of Kazakhstan "Salamatty Kazakhstan" for 2011–2015, approved in November 2010.
(4) Message from the President of the Republic of Kazakhstan N. Nazarbayev. to the people of Kazakhstan on December 14 2012 "Strategy" Kazakhstan-2050 ": a new political course of the held state".
(5) The state program "Information Kazakhstan 2020", approved in January 2013.
(6) The concept of e-health development in the Republic of Kazakhstan for 2013–2020, approved by order in September 2013 and road map.
(7) The State Health Development Program of the Republic of Kazakhstan "Densauly" for 2016–2019, approved in January 2016.

(8) Law of the Republic of Kazakhstan of November 24, 2015 No. 418-V "On Informatization".

(9) Project of the World Bank and the Government of the Republic of Kazakhstan "Transfer of technologies and institutional reform in the health sector of the Republic of Kazakhstan.

List of officially approved regulatory and legal acts that regulate the development of e-health in the Republic of Kazakhstan; a number of important medical automated systems integrated into a single health information system was studied in [2]. A detailed review of the current automated information systems in medical organizations of the country was conducted, using methods of system and structural-logical analysis, SWOT analysis was carried out.

As part of the development of e-health in accordance with the approved Concept of e-Health Development of the Republic of Kazakhstan information and communication technologies will be introduced into the medical sphere everywhere. The focus of e-health will be the formation of a single health information space in which all interested parties, including the patient, have access to the necessary information, regardless of the type of information systems used.

2   Development of e-Health in Kazakhstan

One of the most important strategic directions for the development of the healthcare system is the organization of a single information space and its technological infrastructure.

By today, Kazakhstan's healthcare sector is transitioning to automation of medical information processing and document management. This implies an increase in processing speed, thereby improving the quality of patient care, facilitating the work of medical and medical personnel.

In order to implement e-health in Kazakhstan, it is planned to develop and implement standards that enable the realization of a service-oriented architecture. In turn, it will ensure full interoperability between information systems involved in supporting health processes. That is, this architecture allows the system to interact and function with other products or systems without any restrictions on access and implementation.

Today, the information and technology platform, the so-called Unified Health Information System (UHIMS), is actively implemented in the country, the main purpose of which is to create an information healthcare structure of the Republic of Kazakhstan that corresponds to the level of economic, social, technical and technological development of society and ensures the rational use of health resources in more quality provision of medical services to the population [3].

The creation of UHIMS involves the fulfillment of a number of tasks based on the development and implementation of uniform standards for the exchange of medical data, the use of a unified system for identifying objects of account and subjects of information interaction in health care. It is planned to implement centralized management and open access to a database of common classifiers, directories and standards, including a database of standards of medical care, patient records, state registries of medicines and medical products [4].

Within the framework of the State Health Development Program of the Republic of Kazakhstan "Densaulyk" for 2016–2020, the development of a unified national health information system (hereinafter - UHMIS) will continue to develop common standards, technological specifications and characteristics of various information systems for the required functionality.

It is clear that the task of forming the UHMIS can be solved only at the state level and only for many years. At the same time, the technological prerequisites for its solution already exist. Two private information technologies (in addition to many other components), which have received significant development in recent years, are the basic ones for solving this problem: telemedicine and medical information systems [5].

By the end of 2018, it is planned to create the necessary software and hardware for the implementation of the Electronic Health Passport (hereinafter - EHP), a single repository of analytical health data, an integration bus, tools for maintaining single classifiers, directories and registers.

EHP will become the central link ensuring the interaction of medical information systems and providers of medical services through the implementation of a standardized model of medical information. At the same time, information security mechanisms and protection of personal and confidential data will be provided.

Most medical institutions are currently developing an environment using medical information technology, through the introduction of integrated access to clinical information. Here information technologies and their tools help with administrative and financial issues, in scientific research, in office automation, and also helps patients. At the heart of these evolving integrated environments is an accessible, confidential, secure and integrated electronic medical record [6–9].

3   Health Information Systems: Architectural Models and Management

From a functional point of view, e-health in Kazakhstan supports three main levels of the health system:

Centralized management at the national and regional levels: this includes central planning, resource management, the rules and procedures to be followed, overall financial control, quality and safety control.

Primary health care: this level includes all systems that support services provided to citizens throughout the national or regional territory. It includes all service providers, such as general practitioners, local customs, etc.

Secondary health care: this level refers primarily to systems that support health processes among health care providers.

These three levels are usually interrelated only with respect to administrative and accounting flows, but the potential exchange of data between different layers makes ICTs important for both exchange and processing of large sets of clinical data. These data provide a great potential for the future development of health care. In fact, ICT solutions that are now present in the healthcare industry have the opportunity not only to simplify the relationship between patients and physicians that improve the overall effectiveness of health services - but also to better control the entire health care system.

Functionality and architecture of e-health information systems, as well as the need for standardization is determined by the information model EMR (Electronic Medical Record)/EHP (Electronic Health Passport) (Fig. 1).

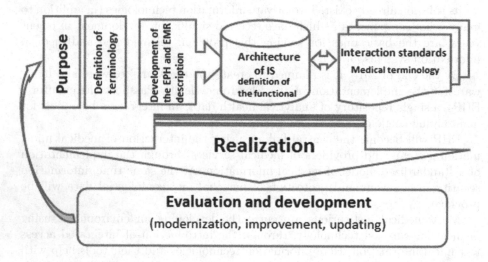

Fig. 1. General scheme for the implementation and development of e-health.

The creation of e-health is based on the following fundamental directions:

- digitization of information flows at the national and regional levels;
- development of a national, as well as a regional social insurance card;
- the presence of an integration bus and a single repository that provides centralized storage of medical and non-medical health data, including electronic health passports of each citizen of the Republic of Kazakhstan (Fig. 2);
- development of regional infrastructure supporting online services for citizens;
- development of a strong set of interrelations between providers of medical services (secondary care) and general practitioners (primary health care);
- establishment of a regional electronic health record, which will subsequently be integrated at the national level;
- digitization of the processes of providing services with secondary care.

Among other priority areas of e-health, the State Program "Information Kazakhstan 2020" noted the following:

- creation of a unified database of medicines;

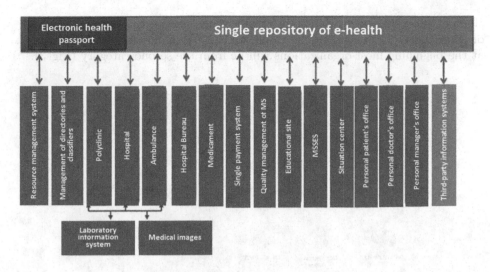

Fig. 2. The proposed conceptual functional architecture of e-health.

- creation of electronic system of medical prescriptions and prescriptions;
- introduction of a system of telemedicine bracelets for monitoring the state of health;
- automation of ambulance services;
- creation of a database of scientific and medical information [3].

Technical architecture is the architecture of the hardware and software infrastructure that ensures the operation of application systems and the execution of operational (non-functional) requirements for the architecture of application systems and information.

Currently, the e-health architecture contains two types of information systems and applications. The first part is developed on the principle of client-server desktop applications. When using these systems, it is assumed that all MOs have servers within the enterprise itself, and all users work with systems on the local network from their workstations. During the night, all the data that was updated during the day is sent from the local server to the top (to the level of the area to ensure data backup), where the data is synchronized, and downward possible changes in directories, data structures and software codes are transmitted. Exchange of operational information necessary for the process of medical care between centralized systems and nodes of UHMIS systems is carried out by accessing services through messaging through the transport environment. The second part of the applications is developed on the web technologies, in which users via web browsers access directly to the central data center of the Ministry of Health of the Republic of Kazakhstan, where web servers, applications and databases are located.

In the future, the transition to cloud technologies is expected. The availability of a backup center for processing medical data (MDC) and a communication channel is considered. In this architecture, there are two categories of medical organizations: (1) working in the cloud, and (2) working with the local server, at the beginning most organizations will be from the second category (Fig. 3).

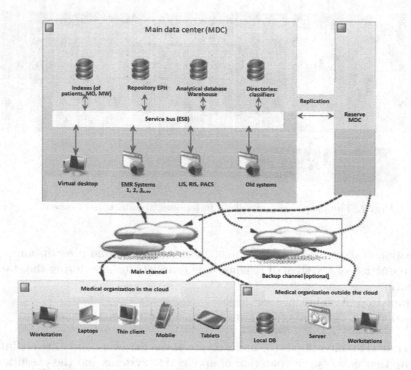

Fig. 3. The architecture of software and hardware in e-health.

Inside the MDC, all databases will be stored, among which the following are highlighted:

- patients Index (PI), which contains the basic identification data for each patient (the personal data are stored in encrypted form in a database), demography and other basic data;
- Electronic health passport, which stores the basic medical data of each patient;
- Specialized databases of individual EMR and web applications, as well as centralized information systems operating in the cloud;
- Data warehouse for statistical calculations.

Data center servers will be treated as existing systems (centralized database from local systems and web applications), and software for virtual desktops and new systems (EHP, IP Polyclinic, IS Hospital, IS Ambulance), etc.

Now the service "Software as a Service" (SaaS) is becoming more widespread. Consequently, cloud computing services are becoming more popular. Often they even completely displace local systems and data stores. It is assumed that such systems can rely solely on SaaS services and provide all the necessary functions through thin clients, which are a combination of inexpensive hardware, an operating system and a web browser.

The health benefits of the SaaS model are undeniable. These advantages are expressed in the following points:

(1) Availability of a single information space. The ability to quickly access computing and information resources when needed is largely the ideal solution for many hospitals, medical clinics and doctor's offices. This should provide them with an opportunity to improve services for their patients.
(2) All system data is stored in a single storage and processed on the system servers. The problem, at least for now, is maintaining patient privacy. The risks of providing sensitive patient data, especially in public cloud infrastructures, continue to constrain the pace of cloud adoption.
(3) Updates and backups are done automatically. The distribution of electronic medical records, implemented on the basis of cloud technologies, allows you to automate not only the registration of the patient, but also to conduct centralized monitoring of the process of providing medical services at a higher level. It also facilitates the transfer of statistical, administrative, financial data in electronic format, which greatly simplifies their processing and optimizes business processes.
(4) System requirements to jobs user is minimal, no need to install and configure, thereby saving time and money.
(5) Possible access from any node of the system. The cloud brings powerful it resources to healthcare professionals, healthcare organizations that are available anywhere in the world. World-class applications and computing infrastructure are available to all without significant upfront investment.

Areas of application of the SaaS model in health care:

- ERP (Enterprise Resource Planning) system that allows you to store and process most of the critical data: workflow, financial report, personnel records, procurement and inventory management;
- System of medical reference information;
- CRM (Customer Relationship Management) - software for organization and automation of work with customers. In health care, these include: management of the patient base, management of the current activities of the health facility);
- Portal solution. In order to raise public awareness, an information portal is being created within the framework of the unified health system. It aims to increase public awareness, transparency and accountability and effectiveness in the health system. Through this, citizens should have access to reliable and complete information about: public health programs, medical services, service providers, medical equipment, medical personnel and their workplace. The portal should allow the citizen to check his/her insurance status and find out what services are available to him/her within the framework of health care programs.

Thus, the introduction of cloud technologies in the medical industry can solve a number of problems associated with the inefficiency of medical services.

Large volumes of various medical data, which are mainly stored in paper form, are difficult to process and analyze. Also this medical data are prevent rapid clinical decision-making.

Although this approach may be rational in some cases, it can not be universal, because it is thus impossible to provide usability, full integration with productive desktop systems, and the ability to work offline (without an Internet connection). Therefore, it is considered expedient to apply the advanced approach "software products and Internet services" (S + S) [10].

4 Conclusion

This article presents the urgency of the process of development of administrative and clinical information systems in the medical sphere of the Republic of Kazakhstan. Particular attention was paid to the introduction, use and features of a unified health information system. Also features of architectural models and management characteristic for e-health of Kazakhstan are described. The functional and architecture of e-health information systems are considered; the proposed conceptual functional architecture of e-health; architecture of software and hardware in e-health.

The introduction of medical information systems, even if well-organized and technologically sound, has its own difficulties. The main problem is that information systems are managed and used by people who have certain beliefs, attitudes and practices, and it will take time to change them.

Most health care providers feel threatened by a system that leads to objective decisions and are suspicious of automation; health consumers believe that more accessible information systems pose a threat to privacy; and there is no close relationship between consumers.

Another important issue of resource management concerns the appropriateness of computerization and at what level. Although the majority of medical institutions in big cities, now have access to the computer equipment, in many rural areas of the country, the computers may still not be available and the computer literacy of the population is not at the proper level.

However, rapidly evolving computer technologies will make health information systems an increasingly effective and powerful management tool for health services. Computer equipment is becoming increasingly available. Software applications for database management and geographic information systems can improve the use of information for decision-making on the health of individuals and society as a whole.

However, the introduction of computer technology is not necessarily the decisive factor that creates efficiency and effectiveness in the health sector. On the other side, the lack of properly trained staff, as well as problems with hardware and software, sometimes lead to the decline and obsolescence of expensive computer equipment without any benefit in decision-making. Therefore, in this area it is also important to train medical staff in the use of information and communication technologies for their professional needs.

In the future, due to the development of e-health, it will be possible to provide medical care to citizens through various telecommunication services, for example, services that provide remote interaction between doctors or doctors and patients, remote monitoring of the patient's health status, maintaining medical records of the patient in electronic form, creating a personal account of the patient.

The development of e-health is considered as a complex project in terms of integration of computing, information and telecommunication infrastructure with the health care system. This will bring the quality of medical care to a new level. E-health technologies will make it possible to monitor the population at a distance, better disseminate information among patients, and improve access to health care.

References

1. Nazarenko, G.I.: Medical information systems: theory and practice. Moscow (2005)
2. Shopabaeva, A.R., Blatov, R.M., Sydykov, S.B., Zhakipbekov, K.S., Elshibekova, K.M.: Information and communication technologies in the healthcare system of the Republic of Kazakhstan: problems and development prospects. Bull. Kazakh Natl Med. Univ. (1), 769–775 (2016)
3. The concept of e-health development of the Republic of Kazakhstan for 2013–2020
4. Kuznetsov, P.P., Stolbtsov, A.P.: Modern information technologies and healthcare development. Med. Almanac (3), 8–12 (2008)
5. Shifrin, M.A.: Creation of a unified health information environment - the mission of medical informatics. Doct. Inf. Technol. (1), 18–21 (2004)
6. Uvalieva, I.M., Belginova, S.A.: Development of the conceptual model of the diagnostic object in the medical diagnosis system. Bull. KazNITI (5), 69–74 (2017)
7. Uvalieva, I.M., Belginova, S.A.: The concept of e-health of the Republic of Kazakhstan. In: EUROPEAN RESEARCH: A Collection of Articles of the XI International Scientific and Practical Conference. Penza, MCSN "Science and Education", pp. 65–69 (2017)
8. Uvalieva, I.M., Belginova, S.A., Ismukhamedova, A.M.: Analysis of decision support systems in medicine. In: ADVANCED SCIENCE: A Collection of Articles of the International Scientific and Practical Conference. At 3 o'clock Part 1. Penza, MCSN "Science and Education", pp. 99–101 (2017)
9. Belginova, S.A., Uvalieva, I.M., Ismukhamedova, A.M.: Use of decision support methods for medical diagnosis. Bull. Shakarim (2)
10. The methodology of building integrated medical information systems, 2nd edn. Microsoft Corporation (2009)

Mathematical Model of Data Processing System for Information Support of Innovative Cluster Works

L. K. Bobrov[✉] and I. P. Medyankina

Novosibirsk State University of Economics and Management, Novosibirsk, Russia
{l.k.bobrov,i.p.medyankina}@edu.nsuem.ru

Abstract. Information systems of innovation management are an important component of cluster organization infrastructure. The creation of such systems involves working with a wide range of commercial information resources. These resources are often used to form thematic databases designed to meet the information needs of cluster organizations. It is highly relevant to find the topology of an information processing system which would minimize overall operational costs for information support of innovative cluster activities. The paper presents the mathematical statement of the problem concerning forming optimal topology of such system. It is shown that finding optimal topology of a distributed information processing system can be reduced to solving the problem of integer linear programming, which would allow minimizing the system operation costs.

Keywords: Mathematical modeling · Innovative clusters ·
Information support · Distributed data processing ·
Optimization models

1 Introduction

The expediency of using various forms of cooperation in organizing information activities was pointed out as early as in the 1990s [1–4]. The interest towards this topic was also provoked by the transition to the market economy in the former Soviet Union countries [5]. Currently, the relevance of this topic is determined by the need to develop innovative activities in conditions of limited financial resources.

Cluster policy is one of effective tools to develop territories and regions [6–9]. Cluster approach is most widely used in the European Union where it

---

This work was supported by a grant from the MES RK (project No. AP05134019 "Development of scientific and methodological foundations and applied aspects of building a distributed information support system for innovation activities, considering the specific features of each of the stages of the innovation life cycle".

Y. Shokin and Z. Shaimardanov (Eds.): CITech 2018, CCIS 998, pp. 64–70, 2019.
https://doi.org/10.1007/978-3-030-12203-4_7

was elevated to the rank of public policy. The Russian Federation has begun to actively support innovative territorial clusters since 2011.

In the Republic of Kazakhstan, the cluster approach is laid down in the State Program of Industrial Innovative Development for 2015-2019. The program provides active financial support for clusters with the highest development potential. They are selected on a competitive basis.

This support is realized in several directions, including the processes of forming supplier bases and creating information platforms. Information systems for innovation management are an important component of cluster organizations infrastructure. The creation of such systems involves working with a wide range of commercial information resources. These resources are often used to form thematic databases designed to meet the information needs of cluster organizations. It is highly relevant to find the topology of such information processing system which would minimize overall operational costs for information support of innovative cluster activities. There are many approaches to solving this class of problems, as well as relevant data processing models, beginning with the simplest set-theoretic models and ending with the most complex simulation models. This article shows that the optimal topology of the information support system for cluster organization can be obtained by means of solving an integer linear programming problem.

2 Mathematical Statement of the Problem

The mathematical model proposed in this article describes the process of creating an information product in terms of cooperation. It also allows solving the problem of optimal distribution of technological operations between members of an innovation cluster in order to achieve the minimum of overall costs for product creation. This model is constructed as follows. Let the analysis of the market and choice of the database subject show that the information flow for database formation is the association of R subjects (rubrics):

$$\Phi = \bigcup_{r=1}^{R} M_r \tag{1}$$

and the volume of each rubric can be estimated by the number of documents belonging to it:

$$|M_r| = V_r.$$

The level of interest of each of the partners $U_1, U_2, \ldots, U_N$ in the processing of each of the R rubrics can be estimated by the vector

$$\bar{p}_n = (p_{n1}, p_{n2}, \ldots, p_{nR}), \quad \left(n = \overline{1, N}\right). \tag{2}$$

In order not only to reflect the participant's interest in processing documents for each of the R rubrics, but also to compare the level of their interest towards any of the rubrics, the following condition must be met:

$$\sum_{r=1}^{R} p_{nr} = 1, \quad (0 \leq p_{nr} \leq 1).$$ (3)

The values p_{nr} can be determined in several ways. In particular, the level of interest of each of the partners in the documents of the rth rubric can be estimated based on the predicted number of queries to each of the M_r arrays from each participant. Then we denote by b_{nr} the number of queries from the partner U_n to the array M_r and get:

$$p_{nr} = b_{nr} / \sum_{r=1}^{R} b_{nr}.$$ (4)

More accurate estimates can be obtained using the notion of completeness of the answer in the database system. Here, as a criterion, the number of documents issued in response to each of the queries is used when searching all M_r arrays. Then, denoting by d_{nr} the total number of documents obtained by searching the array M_r for the entire set of queries B_n of the participant U_n, we have:

$$B_n = \sum_{r=1}^{R} b_{nr}, \quad p_{nr} = d_{nr} / \sum_{r=1}^{R} d_{nr}.$$ (5)

Introducing a system of weighting coefficients $(\beta_{n1}, \beta_{n2}, \ldots, \beta_{nR})$ to consider subjective factors which determine the interest of the participant U_n to the array M_r, we get:

$$p_{nr} = \beta_{nr} d_{nr} / \sum_{r=1}^{R} d_{nr}.$$ (6)

To simplify the time-consuming procedure of formulating a large number of queries, it is possible to use an approach based on the connection between the frequencies of terms in a database and the number of documents issued in response to a query. Then, using the results of a previously organized question-naire of future customers (subscribers), and assuming that each of the queries includes only one term, we get a list of terms (normalized lexical units) for each of the U_n participants:

$$L_n = \left(l_n^1, l_n^2, \ldots, l_n^K \right).$$

Comparing each term with frequency dictionaries of each M_r array, we get:

$$\tilde{d}_{nr} = \sum_{r=1}^{R} F_r,$$

where

$$F_r = \sum_{k=1}^{K} f_{nr}^k,$$

and f_{nr}^k is the frequency of the kth term from the list of partner U_n in the array M_r.

The technological process of information processing by the partners can be represented as an ordered sequence of simple or aggregated operations, the same for any M_r array:

$$O = (O^1, O^2, \ldots, O^Q). \tag{7}$$

Let us denote by t_n^{qi} the volume of unit costs for the ith resource (taking into account the characteristics of software and hardware resources, as well as other factors affecting the real cost of data processing operations in the U_n center) for the operation of O^q by the partner U_n. Also, we consider the fact that the overall cost of ith resource for each of the partners cannot exceed a certain limit value μ_n^i. Taking into account the volumes of the M_r arrays, let us introduce the indicator:

$$\tau_{nr}^{qi} = V_r t_n^{qi}, \tag{8}$$

which characterizes the cost of the ith resource required by the partner U_n to perform the operation O^q on the array M_r. Then the overall cost can be described as:

$$H^1 = \sum_{i=1}^{I} \sum_{n=1}^{N} \sum_{r=1}^{R} \sum_{q=1}^{Q} \omega_{nr}^q \tau_{nr}^{qi}, \tag{9}$$

where

$$\omega_{nr}^q = \begin{cases} 1 - \text{if the } U_n \text{ participant performs the } O^q \text{ operation on the } M_r \text{ array;} \\ 0 - \text{otherwise;} \end{cases}$$

and the equality holds:

$$\sum_{n=1}^{N} \omega_{nr}^q = 1 \quad (q = \overline{1, Q}; \ r = \overline{1, R}). \tag{10}$$

The latter means that each of the O^q operations on any M_r array is necessarily performed, and it is done only by one of the U_n partners, i.e. the principle of one-time processing of information is respected. Thus, there is a problem of minimizing the functional (9) with the equality (10) and limitations held:

$$\omega_{nr}^q = \{0, 1\}; \tag{11}$$

$$\sum_{r=1}^{R} \sum_{q=1}^{Q} \omega_{nr}^q \tau_{nr}^{qi} \leq \mu_i^n. \tag{12}$$

With the solution of this problem, it is possible to find such distribution of work between partners which allows achieving the minimum of the overall costs for creating an information product. However, this is true only if there are no subjective factors affecting the distribution of work between partners, and causing the need for some operations to be performed centrally whereas the solution of others (for example, information service operations themselves) is decentralized. Based on the foregoing, the set of O operations can be divided into three disjoint subsets (O' means centralized operations; O'' means distributed operations; O''' means operations performed by each of the U_n partners), and further focus will be on distributed operations.

We impose a penalty on the execution of operations from O'' in case if the participant U_n (interested in the results of processing documents of the rth rubric) does not perform operations on its processing. Logically, the amount of the penalty should be proportional to the level of interest of the participant and the cost of performing the operation O^q on the array M_r. In this case, the functional (9) will take the following form:

$$H^2 = H_1 + \sum_{i=1}^{I} \sum_{n=1}^{N} \sum_{r=1}^{R} \sum_{q:\, O^q \in O''} (1 - \omega_{nr}^q) p_{nr} \tau_{nr}^{qi}. \tag{13}$$

To simplify the functional, we introduce the notation:

$$\theta_{nr}^q = \sum_{i=1}^{I} \tau_{nr}^{qi}. \tag{14}$$

Then the minimized functional (13) can be rewritten in the form:

$$H^2 = \sum_{n=1}^{N} \sum_{r=1}^{R} \sum_{q:\, O^q \in O''} \omega_{nr}^q \theta_{nr}^q + \sum_{n=1}^{N} \sum_{r=1}^{R} \sum_{q:\, O^q \in O''} (1 - \omega_{nr}^q) p_{nr} \theta_{nr}^q. \tag{15}$$

Thus, the task of obtaining the optimal topology of the information management system for a cluster organization is described by means of the mathematical model (15), (10), (11), (12), which, as you can see, belongs to the class of integer linear programming models.

3   Results and Discussion

The results of numerical calculations for the proposed model allow us to determine the distribution of work between cluster members. This distribution allows to achieve the minimum of overall costs for operating the information support system of this cluster organization with given resource constraints. Figure 1 shows the graphical interpretation of the problem solution results (15) with constraints (10), (11), (12), and obtaining the network topology of distributed information processing in cluster organization conditions. In these conditions, any of the participants performs at least one operation on processing information of documents. There is at least one thematic rubric, and none of them can be duplicated.

Practical testing of the proposed model was carried out during the creation and implementation of an automated scientific and technical information system of the Siberian Branch of the Academy of Sciences. It covers research institutes of

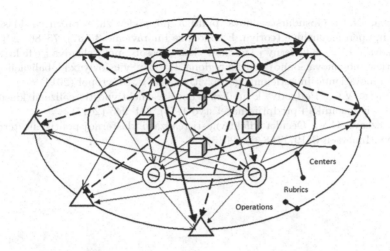

Fig. 1. Graphical interpretation of the results obtained

the Novosibirsk Scientific Center and information files in the fields of chemistry, biology, information sciences, environmental protection, etc. The results obtained allow to reduce financial costs for the creation and operation of the system almost twice [4].

4 Conclusion

The conditions of public-private partnership in solving the development problems of innovation activity cluster forms in the Republic of Kazakhstan involve new technologies of interaction between cluster organizations members. In some cases, this makes it possible to achieve a significant reduction in the cost of creating components of cluster organization infrastructure.

When forming the information infrastructure of a cluster, it is possible to organize a system of distributed information processing which would minimize the total cost of system operation.

References

1. Carpenter, K.H.: Competition? Collaboration? and cost in the new knowledge environment. Collect. Manag. **21**(2), 31–46 (1996)
2. Lesk, M.E.: The organization of digital libraries. Sci. Technol. Libr. **17**(3–4), 9–25 (1999)
3. Angelis, J.: A new look at community connections. III. Libr. **81**(1), 23–24 (1999)
4. Elepov, B.S.: Proektirovanie i e'kspluataciya regional'ny'x ASNTI. Nauka, Novosibirsk (1991)
5. Bobrov, L.K.: Strategicheskoe upravlenie informacionnoj deyatel'nost'yu bibliotek v usloviyax ry'nka. NGAE'iU, Novosibirsk (2003)

6. Anisova, N.A.: Gosudarstvennaya politika po podderzhke razvitiya klasterov: novacii, operezhayushhie teoriyu. E'konomika i upravlenie **3**(137), 75–86 (2017)
7. Novikova, I.V.: Klasterny'j podxod kak sposob povy'sheniya e'ffektivnosti investicionno-innovacionnoj politiki regiona. Regiony' v usloviyax globalizacii: problemy' innovacionno-investicionnogo razvitiya. SKFU, Stavropol (2014)
8. Plastinina, V.G.: Podxody' k analizu i ocenke e'ffektivnosti realizacii klasternoj politiki. E'konomika i predprinimatel'stvo **1**(78), 941–945 (2017)
9. Suroviczkaya, G.V.: Ocenka e'ffektivnosti realizacii klasternoj politiki na territorii regiona. Innovacii **3**(197), 58–60 (2015)

Mechanics-Mathematical Model of Conjugation of a Part of a Trajectory with Conditions of Continuity, Touch and Smoothness

B. O. Bostanov[1]($\boxtimes$), E. S. Temirbekov[2], M. V. Dudkin[3], and A. I. Kim[3]

[1] L.N. Gumilev Eurasian National University, Astana, Kazakhstan
bostanov_bayandy@mail.ru
[2] U.A. Dzholdasbekov IMMash, Astana, Kazakhstan
temirbekove@mail.ru
[3] D. Serikbayev East Kazakhstan State Technical University,
Ust-Kamenogorsk, Kazakhstan
vas_dud@mail.ru, k.a.i.90@mail.ru

Abstract. Development process of combined trajectories, in places of joining conic arcs, undesirable intermittent effects inevitably arise due to the second-order non-smoothness. A second order tangency is considered taking into account the curvature and the equality condition of the arcs curvature radii to be joined at the conjugation points. A kinematic method for determining joints on the basis of a rocker mechanism is given, which ensures smooth joints.

1 Introduction

The scientific interests of the creation and formation of complex trajectories locate in the areas of road construction, aviation industry, shipbuilding, textile production, railway and automobile transport. Combined trajectories are created in the form of conjugate contours, the shapes of which are given by curves of different order and mathematically described by analytical equations. The resulting form should provide an improvement of the functional properties of objects. For example, asymmetric planetary vibration exciters with a combined treadmill (trajectory) are used to improve the performance of road vibratory rollers. Similarly, to improve the aerodynamic properties of the aircraft, combined wing shapes are created, and the smooth geometric shapes of the hull greatly improve the seaworthiness of the vessel. Currently, to implement a smooth transition from one arc to another curve arc, methods of patterns, transformations and second-order curves (conics) are used. These methods provide only smoothness of the first order. The study of conics is due to their wide application in science and engineering practice. These curves are the most important components of the contours of double curvature surfaces. A method of analytic determination of the transition section is proposed to ensure second-order smoothness (smoothness). Mathematical patterns that determine the smoothness of the transition

© Springer Nature Switzerland AG 2019
Y. Shokin and Z. Shaimardanov (Eds.): CITech 2018, CCIS 998, pp. 71–81, 2019.
https://doi.org/10.1007/978-3-030-12203-4_8

are found. The process of finding the starting and ending points of the docking is modelled by the movement of the rocker mechanism [1–8].

The contours of the technical product lines are a combination of lines, which in most cases smoothly passing from one to the other.

A smooth transition of one line to another from a transitional line is called conjugation.

The following methods are used to identify intermediate curvilinear sections: (a) template curve, (b) nonlinear transformations, (c) second-order curves. These methods of constructing the curves are widely used in the design of curvilinear sections of the trajectories [9–13].

However, all these methods are approximate.

The process of determining the position of the finishing point is proposed with the condition that the smoothness ratio be simulated by a rocking mechanism. With the motion of the rocking rock of the rocking mechanism, the distances from the conjugate points to the point of intersection of the tangents change simultaneously, i.e. The changes in the lengths of tangents whose relations satisfy the smoothness conditions. The proposed method makes it possible to visually, quickly and effectively determine the position of the finish point on the circumference and ensure a non-collapsible connection of the conical arcs. Using the method of determining the position of the conjugate points based on the kinematics of the rocking mechanism, it is possible to smoothly join the conical arcs satisfying the conditions of continuity, tangency and equality of curvature and to create on their basis new models of treadmills (trajectories) from conical arcs that allow eliminating unwanted impact effects.

2   Problem Statement

To implement the second order smoothness between curve arcs, it is proposed to insert a transition arc, the model of which is a conic arc (transition conic). The functional purposes of the transition conic are as follows:

- the arc of the transition conic must necessarily pass through the connecting points A and B;
- at the points of joining A and B the first derivatives must be equal (there are common tangents at the points of docking);
- at the joining points A and B the radii of curvature should be equal.

The fulfillment of the first two conditions means the smoothness of the connection, and the addition of the third condition to them ensures a smooth connection.

Let the combined trajectory be formed from arcs of a circle $x^2 + y^2 = a^2$ and an ellipse $\frac{x^2}{a^2} + \frac{y^2}{b^2} = 1$. Choose an arbitrary starting point on an elliptical arc A. We calculate the radius of curvature ρ_A at this point and draw a tangent $L_{A\tau} = 0$. It is required to determine the position of the end point B in order to realize a smooth conjugation of circular and elliptical arcs. Thus, the endpoint B is not arbitrarily selected and should provide functional purposes. To solve the problem of determining the final joining point, it is necessary to determine the mathematical relationships between the elements of the transition conic at the points A and B.

3   Mathematical Preliminaries

Take the two points $A(x_A, y_A)$ and $B(x_B, y_B)$ on the ellipse (Fig. 1) with the curvature radii ρ_A and ρ_B respectively and draw through them the tangents $L_{A\tau}$ and $L_{B\tau}$ with the normals L_{An} and L_{Bn}. Let A_τ, B_τ and A_n, B_n - points of intersection of tangents and normals with the axis Ox, points A_x, B_x - points of the base of perpendiculars, dropped from points A and B on the axis, and Ox, A_h, B_h and A_d, B_d, are the points of the base of perpendiculars dropped from the points A, B and the center of the ellipse O tangent to $L_{B\tau}$ and $L_{A\tau}$. The tangents $L_{A\tau}$ and $L_{B\tau}$ mutually intersect at the point E [23].

By connecting the A, B and E points, we obtain the $\triangle AEB$ triangle. A triangle made up of the tangents $L_{A\tau}$, $L_{B\tau}$, and the chord L_{AB}, and also containing the inside of the arc of the ellipse $\smile AB$ will be the base triangle. EC is the median of the base triangle [23].

We denote by, $n_A = AA_n$, $n_B = BB_n$ - are the lengths of the normals L_{An} and L_{Bn}, $\tau_A = AA_\tau$, $\tau_B = BB_\tau$ - the lengths of the tangents $L_{A\tau}$ and $L_{B\tau}$, $s_A - A_xA_n$, $s_B = B_xB_n$ - are the lengths of the subnormals of the points A and B, $m_A = A_xA_\tau$, $m_B = B_xB_\tau$ - are the lengths of the tangent points A and B, $l_A = AE$, $l_B = BE$- are the lengths of the tangent segments $L_{A\tau}$ and $L_{B\tau}$, prior to their intersection at the point E, $d_A = OA_d$, $d_B = OB_d$ - are the distances of the center O to the tangents $L_{A\tau}$ and $L_{B\tau}$, $h_A = AA_h$, $h_B = BB_h$ - the distances of the points A and B to the tangents L_B and L_A, $\alpha = \angle BAE$, $\beta = \angle ABE$ - are the angles between the tangents $L_{A\tau}$, $L_{B\tau}$ and the chord AB, $\alpha_E = \angle AEC$, $\beta_E = \angle BEC$ - are the angles between the tangents $L_{A\tau}$, $L_{B\tau}$ and the median EC [23, 24].

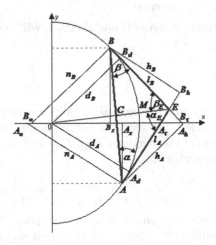

Fig. 1. Basic triangle and its elements

The considered lengths of the segments and the angular quantities will be elements of the basic triangle and find the connecting ratios between them through the radius of curvature [23].

From analytic geometry it is known that the radius of curvature of an ellipse is inversely proportional to the cube of the distance from the center to the tangent at the corresponding point [23]

$$\rho_M = \frac{a^2 b^2}{d_M^3}.$$

We introduce the coefficient defined as the cubic root of the ratio of the radii of curvature [23]:

$$\sqrt[3]{\frac{\rho_A}{\rho_B}} = \frac{d_B}{d_A} = \eta.$$

The coefficient η introduced by us is called the coefficient of curvature [23].

Four points: the center of the ellipse O, the points A_d and B_d of the base of the perpendiculars and the point of E intersection of the tangents are on the same circle [14]. Consider rectangular triangles $\Delta O A_d E$ and $\Delta O B_d E$, the vertices of which lie on the intersection circle and apply the sine theorem [23]. Then

$$\frac{\sin \beta_E}{\sin \alpha_E} = \frac{d_B}{d_A} = \eta.$$

Now consider the triangles ΔACE and ΔBCE, which we get from the basic triangle by dividing the median EC, i.e. $AC = BC$, and similarly applying the sine theorem we obtain [23]

$$\frac{\sin \beta_E}{\sin \alpha_E} = \frac{\sin \beta}{\sin \alpha} = \eta.$$

For a basic triangle, we have [23]

$$\frac{\sin \beta}{\sin \alpha} = \frac{l_A}{l_B} = \eta.$$

From rectangular triangles $\Delta A A_h B$ and $\Delta A B_h B$ with a common hypotenuse AB (chord), we get $\frac{\sin \beta}{\sin \alpha} = \frac{h_A}{h_B} = \eta$.

4   Main Results

Statement. If we have two points A and B an ellipse with radii of curvature ρ_A and ρ_B, then the relationship between the corresponding elements of the basic triangle ΔAEB, made up of the tangents $L_{A\tau}$, $L_{B\tau}$ and the chord L_{AB}, is equal to the smoothness coefficient η [23]:

$$\frac{\sin \beta}{\sin \alpha} = \frac{\sin \beta_E}{\sin \alpha_E} = \frac{d_B}{d_A} = \frac{l_A}{l_B} = \frac{h_A}{h_B} = \frac{n_A}{n_B} = \eta \tag{1}$$

The obtained relations (1), characterizing the properties of the elliptical treadmill, allow determining the position of the point and constructing a transition section [23].

5   Simulation of Connection

The process of finding the position of a point B on a circle that satisfies the $\frac{l_A}{l_B} = \eta$ relation can be modeled by a link mechanism (Fig. 2) [23].

In the rocking mechanism, the guides AE and BE correspond to the directions of the tangents $L_{A\tau}$, $L_{B\tau}$, and the stone E represents the point of their intersection. The movement of the stone leads to a change in the length of the tangents. Value d_A - the distance from the center O to the tangent $L_{A\tau}$, m_A - the distance from the center O to the normal L_{An}, γ - the varying angle of inclination of the tangent $L_{B\tau}$, r - the radius of the connected arc of the circle [23].

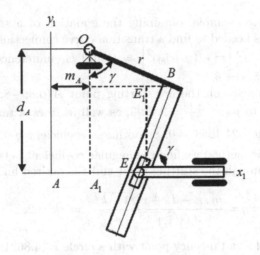

Fig. 2. Rocking mechanism

The equation of motion of the rock of the wings as a function of the angle γ [23]:

$$x_E = (m_A + r\sin\gamma) - \frac{d_A - r\cos\gamma}{tg\gamma}.$$

Similarly, for the point B we have [23]

$$\begin{cases} x_B = m_A + r\sin\gamma \\ y_B = d_A - r\cos\gamma \end{cases}$$

Further, using the dependence $l_A = \eta l_B$, we obtain the equation for determining $k = tg\gamma$ [23]:

$$\frac{m_A + r\sin\gamma + \frac{r}{k\sqrt{1+k^2}} - \frac{d_A}{k}}{m_A^2 + r^2 + d_A^2 + 2r\left(m_A\frac{k}{\sqrt{1+k^2}} - d_A\frac{1}{\sqrt{1+k^2}}\right)} = \eta$$

Applying this method, one can find a conic section having two given tangents at two points with given radii of curvature and passing through the third point in the form of the Lyming equations [15–19, 23]:

$$F(x, y) = (1 - \lambda) L_{A\tau} L_{B\tau} + \lambda L_{AB}^2 = 0.$$

The third point is found from the condition for determining the engineering discriminant for a known radius of curvature [23]

$$f = \frac{1}{1 + \sqrt{\frac{2l_A^2}{h_A \rho_A}}}.$$

Using a concrete example, obtaining the equation of a smooth transition conic is shown. It is needed to find a transition curve connecting lemniskats arcs $\left[x^2 + (y + y_0)^2\right] - 2c^2 \left[x^2 - (y + y_0)^2\right] = 0$ and circumference $x^2 + y^2 = r^2$, where $y_0 = 7$, $c = 5$, $r = 6$.

Choose a lemniscate on the arc starting point $A(6.8; -8.1)$ the radius of curvature is equal to $\rho_A = \frac{2c^2}{3\rho} = 2.4195$, as well as draw a tangent through it $L_{A\tau} = 1.9155x - y - 21.1258 = 0$. Smoothness coefficient $\eta = \sqrt[3]{\frac{\rho_A}{\rho_B}} = 0.7388$. Using the kinematic method we find the angular coefficient between the tangent ones, conducted through the starting point and the desired finishing point B:

$$\eta = \frac{m_A k - d_A + r\sqrt{1 + k^2}}{d_A \sqrt{1 + k^2} - r} \Rightarrow k = 2.$$

Point B defined as a tangency point with a circle $B(4.8621; 3.5157)$ and the tangent equation at the finish point is: $L_{B\tau} = -1.383x - y + 10.2399$. Chord equation $L_{AB} = -5.994x - y + 32.6589 = 0$.

Find the length of the tangent $l_A = 5.859$ and distance $h_A = 5.238$ from the starting point A up to the tangent $L_{B\tau}$, which are necessary to calculate the engineering discriminant $f = \frac{1}{1 + \sqrt{\frac{2l_A^2}{h_A \rho_A}}} = 0.333$.

The coordinates of the point M through which the smooth transitional conic passes is determined by

$$\sigma = \frac{f}{1 - f} = 0.4993$$

$$\begin{cases} x_M = \frac{x_C + \sigma x_E}{1 + \sigma} = 7.0556 \\ y_M = \frac{y_C + \sigma y_E}{1 + \sigma} = -2.498 \end{cases}$$

The coefficient λ in the Lyming equation is calculated by the formula

$$\lambda = \left. \frac{L_{A\tau} L_{B\tau}}{L_{A\tau} L_{B\tau} - L_{AB}^2} \right|_{\substack{x = x_M \\ y = y_M}} = 0.23035.$$

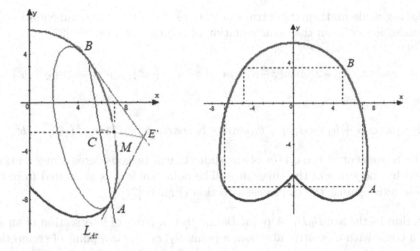

Fig. 3.

Equation of transition conic with continuity, tangency and smoothness conditions

$$F(x,y) = (1 - \lambda)(1.9155x - y - 21.1258)(-1.383x - y + 10.2309) +$$

$$+ \lambda(-5.994x - y + 32.6589)^2 = 0$$

or

$$F(x,y) = 27.0768x^2 + 10.2088xy + 4.3412y^2 - 228.4166x - 28.9158y + 344.4504 = 0.$$

The desired conic is represented as a displaced rotated ellipse [23]

$$\frac{[(x - a_0)\cos\alpha + (y - b_0)\sin\alpha]^2}{a_k^2} + \frac{[-(x - a_0)\sin\alpha + (y - b_0)\cos\alpha]^2}{b_k^2} = 1 \quad (2)$$

where a_k and b_k the semiaxes of the desired conic a_0, b_0 are the coordinates of the center of the conic displacement, and α is the angle of rotation of the focal axis [23].

The movement of the center of the vane of the exciter is described by changing the distance from the center of the runner to the anchoring point of the carrier, i.e. polar radius R. Therefore, all the component curves (circle, ellipse and transition conic) with which we form the roll must be described in polar coordinates (R, φ). The anchoring point of the carrier (the origin) is in the common geometric center of the circular and elliptical parts of the roll [23].

The components - the circular and elliptical parts - of the roller in polar coordinates are described by the equations $R = a$ and $R = \dfrac{b}{\sqrt{1 - e^2 \cos^2 \varphi}}$, respectively, where a and b are the semiaxes of the given ellipse are, $e = \dfrac{a^2 - b^2}{a^2}$ the eccentricity of this ellipse.

Having made mathematical transformations, from the Cartesian equation (2) it is possible to obtain the polar equation of the transition conic [23].
Then

$$R^2[1-e^2\cos^2(\varphi-\alpha)]+2Rq\sin(\varphi-\alpha)-p\cos(\varphi-\alpha)(1-e^2)]+[p^2(1-e^2)+q^2-b^2]=0,$$

where

$$p=a_0\cos\alpha+b_0\sin\alpha,\ q=a_0\sin\alpha-b_0\cos\alpha,\ g=p^2(1-e^2)+q^2-b^2.$$

The movement of the center of the slider C will be considered along sections divided by the joints of the curve arcs. The polar angle φ is measured from the abscissa axis against the clockwise direction (Fig. 3) [23].

1. Section 1 - the arc B_0B_1. A point $B_0(x_0, 0)$ is a point of intersection of an arc of a conic with a positive abscissa, a point $B_1(x_1, y_1)$ is a point of connection of an arc of a conic with an arc of a circle $0 \le x \le x_1$, $0 \le y \le y_1$. The polar angle is $0 \le \varphi \le \varphi_1$ [23].

$$R=\frac{1}{1-e^2\cos^2(\varphi-\alpha)}\{-A\cos(\varphi-\beta)+\sqrt{A^2\cos^2(\varphi-\beta)-g[1-e^2\cos^2(\varphi-\alpha)]}\}$$

where [23]

$$q\sin(\varphi-\alpha)-p(1-e^2)\cos(\varphi-\alpha)=A\cos(\varphi-\beta),$$

$$A=\sqrt{p^2(1-e^2)^2+q^2},\ \cos\beta=\frac{p(1-e^2)}{A},\ \sin\beta=\frac{q}{A}.$$

2. Section 2 - an arc B_1B_2. Point $B_2(x_2, y_2)$- the point of connection of the arc of a circle with an arc of a conic $x_2 \le x \le x_1$ [23], $0 \le y \le y_2$;
The polar angle $\varphi_1 \le \varphi \le \varphi_2$. Equation of motion $R=a$.
3. Section 3 - the arc B_2B_3. The point $B_3(x_3, y_3)$ is the point of joining the arc of the conic with the arc of the ellipse $x_2 \le x \le x_3$, $0 \le y \le y_3$; The polar angle $\varphi_2 \le \varphi \le \varphi_3$ [23].

$$R=\frac{1}{1-e^2\cos^2(\varphi+\alpha)}\{A\cos(\varphi-\beta)+\sqrt{A^2\cos^2(\varphi-\beta)-g[1-e^2\cos^2(\varphi+\alpha)]}\}$$

4. Section 4 - arc B_3B_4. Point $B_4(x_4, y_4)$- the point of connection of the arc of an ellipse with an arc of a conic $x_3 \le x \le x_4$, $0 \le y \le y_4$ [23]; The polar angle $\varphi_3 \le \varphi \le \varphi_4$.

$$R=\frac{b}{\sqrt{1-e^2\cos^2\varphi}}$$

5. Section 5 - the arc B_5B_0 is part of the right conic.

The diagram of a smooth change in the polar coordinate R, describing the movement of the center of the slider of the reclosure on the combined treadmill, is shown in Figs. 4 and 5 [23].

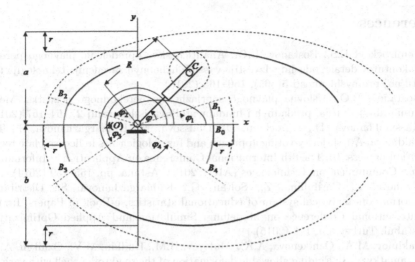

Fig. 4. Combined roll

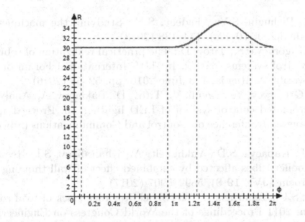

Fig. 5. The diagram of the movement of the center of the runner

6   Conclusion

Thus, the original way of connecting the treadmill of a planetary vibration exciter, obtained by connecting the arc of an ellipse with a circular arc with a radius equal to one of the semi-axes, is obtained. Moreover, the connection point has a common tangent, and does not have a jump along the curvature. The presented method and analytical dependencies of the smooth connection of two curves described by different equations allows making a dynamic calculation of the planetary vibration exciter, the treadmill is described by the intrusion curves (conic). The result is a smooth connection of treadmill sections, which ensures that the inertial runner of the planetary exciter moves along it evenly without drops and jumps at the junction points of various curves.

References

1. Temirbekov, E.S., Bostanov, B.O.: Analiticheskoe opredelenie plavnogo perehodakonturov detalej odezhdy. Izvestija vysshih uchebnyh zavedenij. Tehnologija tekstil'noj promyshlennosti **5**(365), 160–165 (2016)
2. Bostanov, B.O.: Uslovija plavnogo soprjazhenija perehodnogo uchastka. Mezhdunarodnyj zhurnal prikladnyh i fundamental'nyh issledovanij **2**, 164–167 (2016)
3. Kassymkhanova, D., Kurochkin, D., Dinissova, N., Kumargazhanova, S., Tlebaldinova, A.: Majority voting approach and fuzzy logic rules in license plate recognition process. In: The 8th International Conference on Application of Information and Communication Technologies (AICT 2014), Astana, pp. 155–159 (2014)
4. Uvalieva, I., Garifullina, Z., Soltan, G., Kumargazhanova, S.: Distributed information-analytical system of educational statistics, e-Book of Papers. In: 6th International Conference on Modelling, Simulation and Applied Optimization, Istabul, Turkey, pp. 1–6 (2015)
5. Sakimov, M.A., Ozhikenova, A.K., Abdeyev, B.M., Dudkin, M.V., Ozhiken, A.K., Azamatkyzy, S.: Finding allowable deformation of the road roller shell with variable curvature. News Natl. Acad. Sci. Repub. Kaz. Ser. Geol. Tech. Sci. **3**(429), 197–207 (2018)
6. Doudkin, M.V., Pichugin, S.Yu., Fadeev, S.N.: Studying the machines for road maintenance. Life Sci. J. **10**(12s), 134–138 (2013)
7. Baklanov, A., Grigoryeva, S., Titov, D.: The practical realization of robustness for LED lighting control systems, part 2. In: 11th International Forum on Strategic Technology, Novosibirsk, Russia, 1–3 June 2016, pp. 52–57 (2016)
8. Grigoryeva, S., Grigoryev, Ye., Sayun, V., Titov, D., Baklanov, A.: Analysis energy efficiency of automated control system of LED lighting. In: Proceedings of 2017 International Siberian Conference on Control and Communications (Sibcon). IEEE (2017)
9. Bergander, M.J., Kapaeva, S.D., Vakhguelt, A., Khairaliyev, S.I.: Remaining life assessment for boiler tubes affected by combined effect of wall thinning and overheating. J. Vibroeng. JVE **19**(8), 5892–5907 (2017)
10. Dzholdasbekov, S.U., Temirbekov, Y.S.: Shock-free race track of road roller vibration exciters. In: 2011 Proceedings of the World Congress on Engineering, WCE 2011, London, UK, 6–8 July 2011, vol. III (2011)
11. Rybakova, D.A., Sygynganova, I.K., Kumargazhanova, S.K., Baklanov, A.E., Shvets, O.Y.: Application of a CPU streaming technology to work of the computer with data coming from the network on the example of a heating station. In: International Conference of Young Specialists on Micro/Nanotechnologies and Electron Devices, EDM, pp. 128–130 (2017D)
12. Kumargazhanova, S., et al.: Development of the information and analytical system in the control of management of university scientific and educational activities. Acta Polytech. Hung. **4**(15), 27–44 (2018)
13. Baklanov, A., Grigoryeva, S., Gyorok, G.: Control of LED lighting equipment with robustness elements. Acta Polytech. Hung. **13**(5), 105–119 (2016)
14. Kumargazhanova, S., Erulanova, A., Soltan, G., Suleimenova, L., Zhomartkyzy, G.: System of indicators for monitoring the activities of an educational institution. In: Ural Symposium on Biomedical Engineering, Radioelectronics and Information Technology (USBEREIT), pp. 179–182 (2018)
15. Surashev, N., Dudkin, M.V., Yelemes, D., Kalieva, A.: The planetary vibroexciter with elliptic inner race. Adv. Mater. Res. **694–697**, 229–232 (2013)

16. Kim, A., Doudkin, M.V., Vavilov, A., Guryanov, G.: New vibroscreen with additional feed elements. Arch. Civ. Mech. Eng. **17**(4), 786–794 (2017)
17. Gabdyssalyk, R., Lopukhov, Y.I., Dudkin, M.V.: Study of the structure and properties of the metal of 10Cr17Ni8Si5Mn2Ti grade during cladding in a protective atmosphere. News Natl. Acad. Sci. Repub. Kaz. Ser. Geol. Tech. Sci. **2**(428), 95–103 (2018)
18. Doudkin, M.V., Pichugin, S.Yu., Fadeev, S.N.: Contact force calculation of the machine operational point. Life Sci. J. **10**(10s), 246–250 (2013)
19. Stryczek, J., Banas, M., Krawczyk, J., Marciniak, L., Stryczek, P.: The fluid power elements and systems made of plastics. Procedia Eng. **176**, 600–609 (2017)
20. Bojko, A., Fedotov, A.I., Khalezov, W.P., Mlynczak, M.: Analysis of brake testing methods in vehicle safety. In: Safety and Reliability: Methodology and Applications. Proceedings of the European Safety and Reliability Conference, ESREL (2014)
21. Doudkin, M.V., Vavilov, A.V., Pichugin, S.Yu., Fadeev, S.N.: Calculation of the interaction of working body of road machine with the surface. Life Sci. J. **10**(12s), 832–837 (2013)
22. Doudkin, M.V., Pichugin, S.Yu., Fadeev, S.N.: The analysis of road machine working elements parameters. World Appl. Sci. J. **23**(2), 151–158 (2013)
23. Bostanov, B.O., Temirbekov, E.S., Matin, D.T.: The model of a transition region with smoothness conditions of the second order. In: International Conference on Analysis and Applied Mathematics (ICAAM 2018) AIP Conference on Proceedings of 1997, pp. 020038-1–020038-6. Published by AIP Publishing (2018)
24. Temirbekov, E.S., Bostanov, B.O., Dudkin, M.V., Kaimov, S.T., Kaimov, A.T.: Combined trajectory of continuous curvature. In: Carbone, G., Gasparetto, A. (eds.) IFToMM ITALY 2018. MMS, vol. 68, pp. 12–19. Springer, Cham (2019). https://doi.org/10.1007/978-3-030-03320-0_2
25. Bostanov, B.O., Temirbekov, E.S., Dudkin, M.V.: Planetarnyj vibrovozbuditel s ellipticheskoj dorozhkoj. In: Proceedings of the All-Russian Scientic and Technical Conference "Rolmehaniki v sozdanii effektivnyh materialov, konstrukcij i mashin HHI veka", 6–7 December 2006, pp. 66–70 . Omsk, SibADI (2006)

Computer Modeling Application
for Analysis of Stress-Strain State
of Vibroscreen Feed Elements
by Finite Elements Method

M. Doudkin[1]([⊠]), A. Kim[1], V. Kim[1], M. Mlynczak[2], and G. Kustarev[3]

[1] D. Serikbayev East Kazakhstan State Technical University,
19, Serikbayev Str., Ust-Kamenogorsk, Kazakhstan
vas_dud@mail.ru, k.a.i.90@mail.ru, k-v-a@list.ru
[2] Wroclaw University of Science and Technology,
27, Wybrzeze Wyspianskiego, 50-370 Wroclaw, Poland
marek.mlynczak@pwr.edu.pl
[3] Moscow Automobile and Road Construction State Technical University,
64, Leningradsky Prospect, Moscow, Russian Federation
proektdm@mail.ru

Abstract. The article presents a three-dimensional solid-state computational model, as well as the results of stress-strain state analysis of
feed elements rods of vibroscreen. An algorithm for solving the problem
numerically using the finite element method is proposed. The obtained
results were used at the designing stage of the platform with feed elements for industrial vibroscreen and subsequently confirmed in work in
a real experiment. The stress-strain state of feed elements rods was analyzed for various bulk materials, conditionally designated A and B, sorted
by vibrating screen, where feed elements were mounted. These materials, in screening process with varying strength, acted on feed elements
rods, the parameters of which did not change for the flow of various bulk
materials. Rods perceived this load pressing in different ways, which was
shown by the finite element analysis.

Keywords: Vibroscreen · Feed elements · Analysis ·
Finite elements method · Modeling

1 Introduction

The proposed article is in addition to the well-known publication "New vibroscreen with additional feed elements" [1], which describes the mathematical and
physical models of the experimental installation of a new vibroscreen with additional feed elements. Feed elements are an additional construction that significantly intensifies the screening process and at the same time the weakest part of
the vibroscreen. Based on the developed mathematical screening model using a

Y. Shokin and Z. Shaimardanov (Eds.): CITech 2018, CCIS 998, pp. 82–96, 2019.
https://doi.org/10.1007/978-3-030-12203-4_9

main platform of feed elements, numerical studies have been carried out to identify the influence of the process parameters on the screening kinetics and the state of various passage particles in the screened layer. The numerical studies shown that the introduction of additional feed elements into the screened material flow leads to an increase in the passing intensity of the bulk material lower class to the sieve, regardless of the size fractions of different bulk materials.

Figure 1 schematically shows a vibrator with additional feed elements (FE) 7 mounted on the frame 6.

During the screening process, the sorted material flow 2 moves along the screen 1 and comes into contact with the feed elements (FE) 7 located along and across the screen. Since the feed elements themselves are installed with a certain step, the material not only rests against them, but also flows around the sides, exerting pressure on them with a certain force for each material.

The collision of bulk material particles with FE leads to the creation of additional chaotic movements of particles in the upper layers of bulk material, either coinciding or different from the forced oscillations direction from the vibration exciter 5 (Fig. 1). While the difference in particles velocities of the particulate material relative to each other, which activates the passing particles process of the bottom fraction to the screen surface through the entire layer of bulk material through forced mixing.

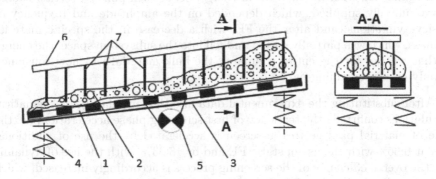

Fig. 1. Scheme of vibroscreen with additional feed elements: 1 - screen, 2 - bulk material, 3 - box, 4 - elastic supports, 5 - vibration engine, 6 - frame, 7 - feed elements.

The productivity of screening before FE can be described by the formula [1]:

$$Q = \mu_{FE} \cdot (S - S_{FE}) \cdot$$

$$\sqrt{\frac{2}{\gamma - \Delta\gamma} \left(\frac{m(t) \cdot (0.02 \cdot A \cdot \vartheta^2)}{S} - \frac{k_F \cdot \gamma(t) \cdot c \cdot (2.22 \cdot A \cdot \vartheta^2)}{2} \right)} \quad (1)$$

It can be seen from the formula (1) that the increase in the productivity of the screening process is possible due to decrease in the specific material mass on the screen or by more intensive mixing of the bulk material on the screen, which leads to slowing down of the material flow, as well as its redistribution and longer residence of the material in the active sieving zone. As a result, the specific weight of the finished product increases, the total amount of material decreases, while the productivity Q increases; because smaller amount of material is distributed over the same area of material flow and experiences less resistance.

Since active screening occurs mainly up to FE, after them the vibrating screen can be considered as a vibration transporter, since the screening process after FE itself is very insignificant, so that the productivity of the process after FE is directly and primarily related to the speed of material transportation along the screen Q = v (t), and velocity of the material passing through the screen is the main researched parameter after the FE [1–6].

$$\frac{\nu_2^2}{\nu^2} = \frac{2m \cdot A_2 \cdot \vartheta^2 \cdot \phi_2^2}{S_2 \cdot (\gamma - \Delta\gamma)} \cdot \frac{S \cdot \gamma(t)}{2m \cdot A \cdot \vartheta^2 \cdot \phi^2} \Rightarrow \frac{\nu_2^2}{\nu^2} = \frac{A_2 \cdot \phi_2^2 \cdot \gamma \cdot S}{A \cdot \phi^2 \cdot (\gamma - \gamma_2) \cdot S_2}. \quad (2)$$

According to the formula (2), it can be concluded that the rate of bulk material after FE should increase for the following reasons:

- due to the decrease in the material resistance to movement, because, after the FE, the number of material pieces decreases per area, respectively, the pieces of the remaining material less interfere with each other;
- in connection with the increase in energy, i.e. at each point, a certain energy was initially supplied, which depended on the amplitude and frequency of sieve vibrations, and after the FE, with a decrease in the specific material mass, which was partially sorted and fell into the subsystem space, the energy that was set for the entire volume of the bulk material was now consumed only on his remaining part.

After substituting the experimental data into the formula (2), mathematical calculations confirmed the tendency of first screening phase acceleration, i.e. the time of material passage to the screen is accelerated by the use of additional FEs at 6.66% with the use of static FE and by 13.33%, with the use of dynamic FE, the overall efficiency of the screening process is accordingly increased, which is graphically depicted in Fig. 2, each state of a vibrating screen (a - without a FE; b - with static RE; c - with dynamic FE) as an increase in the upper horizontal coordinate shows the acceleration of the first screening phase for 6 cells, conventionally accepted lying on the screen.

To carry out experimental studies and analyze the behavior of particles on a natural flat screen, a physical model of vibrating screen was developed and manufactured, which contains a sieve box mounted on elastic supports, vibrator and installed feed elements (FE) above the surface of the screen, rods attached to the frame, while the feed elements can be movable and equipped with a drive, and also fixed removable [7–12].

These requirements, even before the production of feed elements complex, lead to the need for setting new tasks and developing new modeling techniques at the stage of feed elements designing for vibration screens that most fully and adequately take into account the geometry, internal composition and mechanical features of the elements being developed for operation under real operating conditions.

From these conditions, higher requirements to the rigidity and strength of the feed elements arise, while simultaneously striving to reduce their mass, optimality and versatility of the design, while maintaining the reliability of the work and the FE, and the vibroscreen itself.

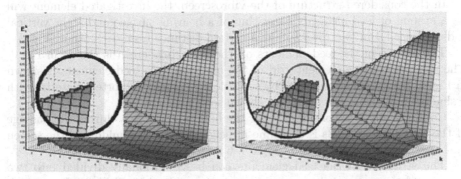

Fig. 2. Graphic image of the acceleration of the first screening phase of the conditional 6th cell: (a) conventional vibroscreen (without FEs); (b) vibroscreen with static FEs; (c) vibroscreen with dynamic FEs, according to the developed mathematical model.

2 FEM Analysis

One of the main methods that reduce the structure development time, which increasing requirements for rigidity and strength characteristics, is the finite element method (FEM), which makes it possible to foresee the behavior and reliability of the experimental design, even before industrial manufacture. This makes it possible to shorten the cycles of experimental studies, modification of the design, processing of design and technological documentation, significant reduction in the cost of production preparation.

Therefore, the analysis of design decisions in the early design stages with the help of simulation systems, including the finite element analysis, which allows modeling the future design and the processes of its testing for various effects at the design stage, acquires special significance.

In this article, the finite element method (FEM) is used to solve the problems of determining the stress-strain state of feed elements rods fixed on the platform, for which we will perform the necessary preliminary graphic constructions, calculations and selection of initial data.

A solid 3D model of a vibrating screen with a separate bearing surface of feed elements rods (Fig. 3) and separate calculation models (Figs. 4, 5, 6 and 7) of feed elements consisting of metal rods mounted on a bearing platform from metal corners were developed and implemented in a system of parametric solid modeling KOMPAS-3DV17.1. The calculations were made in the APM FEM system, which is an integrated tool in the KOMPAS-3DV17.1 for the preparation and subsequent finite element analysis of a three-dimensional solid model (Certification passport No. 330 dated April 18, 2013, issued by the Federal Service for Environmental, Technological and Nuclear Supervision (Rostekhnadzor), FBU "STC NRS"). The calculating core of the APM FEM system for KOMPAS-3D is the "Finite element program system APM Structure 3D [13–17].

In the considered structure of the vibroscreen, the investigated element will be a system of feed elements consisting of a bearing surface and eight vertical rods arranged 4 in two rows and at a certain distance from each other (Fig. 3).

The setting of the feed elements on the vibrating screen is shown in Fig. 3. The calculated finite element grid of the platform with feed elements is shown in Fig. 4. Between feed elements by conditional groups, the sorted material with a shadow trace from the motion is shown schematically.

Conditionally it is considered that the bulk material flow during the passage of the feed elements rods is the same in height both in the middle of the screen and at edges, which is shown in Figs. 5, 6 and 7.

The structure of the feed elements can be conditionally divided into two groups of homogeneous elements - a metal corner (Steel 20, Table 1), from which the bearing platform is made (Corner profile d63x63x5 GOST 8509-93, transverse length $L = 1066$ mm and longitudinal length $L = 340$ mm), which serves as the basis for fastening the rods, and the rods themselves (Steel 45, GOST 1050-2013), mounted on this site.

The behavior of feed elements is simulated in contact with two different bulk materials "A" and "B", which have significant property differences. For each distributed force acting on feed elements, the corresponding objects are selected and the following load parameters are applied: Force vector: for material "A" - $X = 400$; $Y = 0$; $z = 0$; for material "B" - $X = 1200$; $Y = 0$; $z = 0$. Accordingly, the magnitude of the flow force of the bulk material acting on the rods of the feed elements: for material "A" - 400 N, for material "B" - 1200 N.

Accordingly, information on fixations and coinciding surfaces is included in the program, and finite-element grid consisting of 10 nodal tetrahedrons is chosen. It should be noted that tetrahedrons are the only elements that can be used to thicken the adaptive grid. When the density of the grid is increased, the solution becomes more precise. So it can unerringly assume relatively slow change in the stresses in these areas. In these calculations, the adaptive grid is unchanged. For all accepted parameters, static calculations were performed, the results of which are presented in Figs. 4, 5, 6 and 7.

The initial data for solving the problem are summarized in Tables 1, 2 and 3.

The equivalent Mises stress (Fig. 5), SVM (MPa), ranged from the minimum value (0) to the maximum (118,579,194) for bulk material "A" and from 0 to 474.317501 for material "B".

a   b

Fig. 3. Vibroscreen with feed elements: a - general view of vibrating screen with FE; b - feed elements.

Table 1. Data of platform material (Steel 20, GOST 1050-2013).

1	Yield stress (MPa)	235
2	The modulus of normal elasticity (MPa)	200 000
3	Poisson's ratio	0.3
4	Density (kg/m³)	7 800
5	The temperature coefficient of linear expansion (1/C)	0.000012
6	Thermal conductivity (W/m C)	55
7	Compressive strength (MPa)	410
8	Tensile strength (MPa)	209
9	Torsional fatigue strength (MPa)	139

Table 2. Data of feed elements material (Steel 45, GOST 1050-2013).

1	Yield stress (MPa)	560
2	The modulus of normal elasticity (MPa)	210 000
3	Poisson's ratio	0.3
4	Density (kg/m³)	7 810
5	The temperature coefficient of linear expansion (1/C)	0.000013
6	Thermal conductivity (W/m C)	47
7	Compressive strength (MPa)	600
8	Tensile strength (MPa)	294
9	Torsional fatigue strength (MPa)	150

Stress analysis shows their growth for material "B", which is very clearly seen when comparing the FE in Fig. 6, where the middle of the area working with material "B" acquired tints of stresses corresponding to high numerical values of the color scale (Table 4).

Table 3. Parameters and results of feed element location.

	Description	Value
1	Element type	10-nodal tetrahedrons
2	Maximum length of the element side (mm)	10
3	The maximum coefficient of condensation on the surface	5
4	The coefficient of vacuum in the volume	3.5
5	Number of finite elements	114953
6	Number of nodes	235487

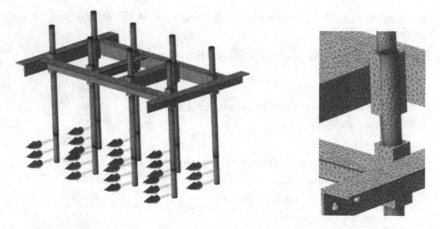

Fig. 4. Finite element grid of the FE construction (a) and its fragment enlarged (b), consisting of tetrahedron grid of finite elements.

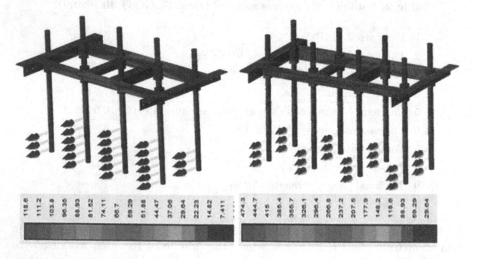

Fig. 5. Equivalent Mises stress of feed elements platform for various bulk materials.

Accordingly, the total linear displacement, USUM (mm), also varied from (0) to a maximum of 3.905783 for bulk material "A" and from 0 to 15.623117 for material "B" (Fig. 6).

As a result of the numerical analysis of the researched system, a picture of the safety factor distribution by structural elements was obtained, expressed by the range of its change. The obtained values of the safety factor ranged from 3.457605 to 1000 for bulk material "A" and from 0.8644 to 1000 for material "B" (Fig. 7). As can be seen from the results of the analysis, the loading regime of the structure for material "A" is safe, as indicated by the value of the minimum safety factor of 3.5. And for material "B", the minimum safety factor is 0.8, which is much less than 3.5. Consequently, the load from the material "A" practically does not affect the efficiency of the FE. The simulated design demonstrated a large safety margin, confirmed by the resulting safety factor.

The scale of the FE image shown in Fig. 7 does not allow us to clearly compare the color variations of the finite element grid, which confirm changes in the calculated values of the coefficient for the loads considered from the action of various materials, but the color gradient decoding below each of the figures shows numerically a significant difference in the indices strength.

Based on the results of modeling and calculation, there were no large deformations or stresses in the structure exceeding the yield strength of the material

Table 4. Inertial characteristics of FE models.

	Name	Value
1	Mass of the model (kg)	55.81
2	Center of the model gravity (m)	(0.000011; −0.117649; −0.000087)
3	The inertia moment of the model relative to the mass center (kg m^2)	(8.229942; 7.831576; 3.248618)
4	The reactive moment with respect to the mass center (N m)	For bulk material "A" (0.656308; −47.820212; −1151.888135)
		For bulk material "B" (2.639079; -191.280431; -4607.557001)
5	The total reaction of supports (N)	For bulk material "A" (−3057.227214; 0.157531; −1.370206)
		For bulk material "B" (−12228.908843; 0.630143; −5.480826)
6	The absolute value of the reaction (N)	For bulk material "A" 3057.227525
		For bulk material "B" 12228.910087
7	The absolute value of the moment (N m)	For bulk material "A" 1152.880514
		For bulk material "B" 4611.526502

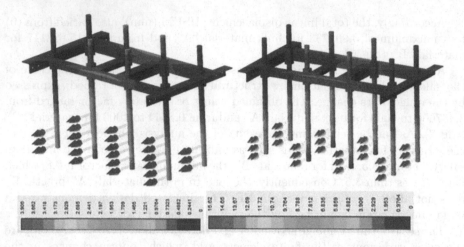

Fig. 6. Distribution of the total linear displacement on the platform. (Color figure online)

when the two materials "A" and "B" were screened, so there was no point in re-solving the problem with the new data.

An exception can be given by a material with special properties that increases the force of the pressure on feed elements an order of magnitude because of its increased density or the adhesion coefficient. The above calculations and modeling were made for the average statistical and widely used materials in the construction. The finite element method gave an approximate, nocturnal solution for the physically and geometrically linear rods of feed elements of the screen, greatly simplifying the final system of equations, determining and showing the reliability of the chosen design.

3   Experiment

The main test of any calculation results is physical experiment. Considering that the above simulation and calculations represent only feed elements simulation of the real design and how accurate the model and mathematical apparatus implementing this model depend on the experimental verification results. Therefore, on the results of the finite element analysis, a natural, industrially applicable platform with feed elements was installed (Fig. 8). The vibroengine is mounted on the platform with the possibility of changing the amplitude and frequency oscillation, tested for durability and reliability in actual production conditions. A comparison was made between the calculated and actual screen weediness, the dependence of sieve weediness on feed elements number, the sieve weediness dependence on the number of FEs rows, the actual and calculated screen weediness was analyzed, by varying the sieve oscillation frequency.

In the experiment, material 2 moving along the sieve 1, under the influence of vibrations generated by the vibration engine 5, is divided into upper and

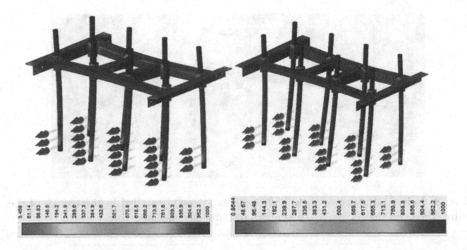

Fig. 7. Distribution of the safety factor for strength on the platform. (Color figure online)

lower fractions. The elastic supports 4 provide the mobility of the box with the screen 1 for generating oscillatory movements. The vibroengine works at a constant speed. Particles of bulk material 2 moving along the screen 1 run into feed elements 7 fixed to the frame 6, which leads to the creation of additional chaotic movements of particles in the upper layers of bulk material 2, either coinciding or differing from the direction of the forced oscillations reported by the engine 5.

This increases the difference in the speed of bulk material particles relative to each other, which activates the process of passing particles of lower fraction to the screen surface through the entire layer of bulk material, which increases the screening process efficiency, as a result, increases the screening capacity while maintaining quality sorting.

To determine the efficiency of the FEs operation, it is necessary to identify the most efficient operating mode without FE. For this purpose the following parameters varied: the amplitude of the sieve A oscillation (0.75–3.0 mm) and the vibration frequency v (8–50 Hz).

After revealing the most rational operating mode of the vibration screen without FE, it is necessary to determine the most effective operating mode of the screen with FE, for this purpose the following parameters were changed: the number of FEs in a row from 1 to 10; the number of FEs rows from 1 to 5.

In this case, the EE are installed at a distance of 200 mm from the beginning of the screening and at a height of 10 mm from the screen, these parameters are established in preliminary experimental studies.

The experiment results on determining the influence of the FE number on the screen weed are given in Table 5, and the graph of the sieve weed dependence on the number of FEs is shown (Fig. 9).

Fig. 8. Industrial platform with feed elements (1 row), installed in the trough of the vibroscreen on the basis of data analysis of finite element modeling.

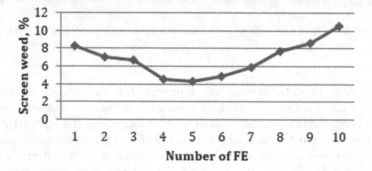

Fig. 9. Graph of the sieve weediness dependence on the number of feed elements.

Table 5. Test results with different number of FE.

FEs number	Weight, g		Screening time (min)	Weediness of the upper class ZB, %
	Above the sieve	Under the sieve		
1	15 112	6 378	1.40	8.31
2	15 000	6 340	1.38	7.01
3	14 650	6 150	1.40	6.62
4	14 770	6 280	1.36	4.55
5	14 830	6 240	1.36	4.31
6	14 770	6 280	1.36	4.91
7	14 650	6 150	1.34	5.92
8	14 760	6 310	1.34	7.71
9	14 760	6 310	1.33	8.59
10	15 100	6 230	1.32	10.56

Table 6. Test results with different number of FEs rows.

FEs number	Weight, g		Screening time (min)	Weediness of the upper class ZB, %
	Above the sieve	Under the sieve		
1	14 830	6 240	1.36	4.31
2	14 530	6 000	1.36	3.5
3	14 550	6 001	1.36	3.51
4	14 530	6 002	1.36	3.52
5	14 489	6 590	1.36	3.79
6	14 552	6 590	1.36	4.52
7	14 466	6 590	1.36	4.78
8	14 588	6 587	1.37	5.59
9	14 003	6 004	1.35	6.98
10	14 554	6 001	1.35	7.99

Further, tests are carried out with a variation of the FE series (Table 6) and an experimental dependence is illustrated (Fig. 10).

On the graph (Fig. 10) the following processes are shown: the lowest index of contamination is reached with the number of rows - from 2 to 10, thus, with the same indices of the minimum level of contamination, a decision is made for further experiments of 2 rows of FE with 5 rods in each row.

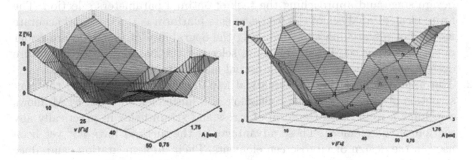

Fig. 10. Graph of the screening weed dependence against the frequency of sieve oscillations, the amplitude of sieve vibrations and the number of FEs, under different inclination angles of the coordinate axes.

The evaluation of the strength and rigidity characteristics of the structure (stresses, deformations, displacement of various points in the structure) by finite elemental analysis is fully confirmed in industrial design tests.

To represent the experiment results, multidimensional graph is shown (Fig. 10); the same 3D image of the screening weed, for greater clarity, is shown at different angles of inclination.

The results of experimental studies show a reduction in the level of the upper fraction contamination up to 3.5%, which confirms an increase in the productivity of the screening process, with the same energy consumption, which leads to the conclusion that it is decreasing.

Thus, according to the experimental results, the most effective mode of the vibroscreening process with a 3.5% weed was detected, with an amplitude of sieve vibrations A = 2 mm, a sieve vibrational frequency v = 25 Hz, a number of FEs in the FEe series of 5 and the number of FEs rows equal to N = 5, the time of material passage along the screen was accelerated to t = 1.36 min. Consequently, with a normative screen weed of 5%, the productivity of the screening process increased by 2.1%, and the process time decreased by 4.4%.

Experimental studies confirmed the theoretical premise that the additional material stimulation on the screen, when compared with the standard screening on a flat screen, provides an increase in the productivity of the screening process by 2.1%, a decrease in the material passage time through the screen by 4.4%, while maintaining normative weed of the material is 5%.

The practical significance of the work is confirmed by the using of prototypes of the vibrating screen with additional feed elements in LLP "CS RNP" in Ust-Kamenogorsk in 2017.

4   Conclusion

1. The simulation model with the finite element method, in contrast to full-scale manufacturing, allows determining the weak places in the design at the design stage and approaching the task of optimal parameters selection. The finite element analysis of the feed elements platform is the best and accurate method of researching and predicting the operability of the structure under given operating conditions, allowing selecting the parameters of the future design reasonably prior to its industrial manufacture.

2. The application of the mathematical apparatus (FEM) simplifies the construction of an object model consisting of a finite elements set. FEM allows obtaining a solution in the form of stress and strain fields in practically any section of the element. These advantages of the method have not yet been used in the design of vibroscreen elements. Their implementation can reduce the metal equipment consumption, increase the reliability of its operation and reduce self-cost and, ultimately, improve the quality of the sorted material.

3. Computer simulation technology with the help of FEM allows reliable determination of the real operational characteristics of products, helps customers to ensure that their products comply with the necessary requirements and standards.

4. The results of experimental studies of FEs, produced according to the results of the FEM analysis, showed a decrease in the level of sieve weed up to 3.5%, which allows increasing the productivity of the screening process and also reducing the energy consumption.

5. Experimental studies confirmed the theoretical premise that the additional material stimulation on the screen, when compared with the standard screening on a flat screen, provides an increase in the productivity of the screening process by 2.1%, a decrease in the material passage time through the screen by 4.4%, while maintaining normative weed of the material is 5%.

References

1. Kim, A., Doudkin, M., Vavilov, A., Guryanov, G.: New vibroscreen with additional feed elements. Arch. Civ. Mech. Eng. **17**(4), 786–794 (2017). https://doi.org/10. 1016/j.acme.2017.02.009
2. Stryczek, J., Banas, M., Krawczyk, J., Marciniak, L., Stryczek, P.: The fluid power elements and systems made of plastics. Procedia Engineering **176**, 600–609 (2017). www.elsevier.com/locate/procedia
3. Doudkin, M., Pichugin, S., Fadeev, S.: Contact force calculation of the machine operational point. Life Sci. J. **10**(39), 246–250 (2013). http://www.lifescience site.com
4. Doudkin, M., Vavilov, A., Pichugin, S., Fadeev, S.: Calculation of the interaction of working body of road machine with the surface. Life Sci. J. **133**, 832–837 (2013). http://www.lifesciencesite.com
5. Giel, R., Mlynczak, M., Plewa, M.: Evaluation method of the waste processing system operation. In: Risk, Reliability and Safety: Innovating Theory and Practice - Proceedings of the 26th European Safety and Reliability Conference, ESREL (2016)
6. Fedotov, A.I., Młyńczak, M.: Simulation and experimental analysis of quality control of vehicle brake systems using flat plate tester. In: Zamojski, W., Mazurkiewicz, J., Sugier, J., Walkowiak, T., Kacprzyk, J. (eds.) Dependability Engineering and Complex Systems. AISC, vol. 470, pp. 135–146. Springer, Cham (2016). https:// doi.org/10.1007/978-3-319-39639-2_12
7. Bojko, A., Fedotov, A.I., Khalezov, W.P., Mlynczak, M.: Analysis of brake testing methods in vehicle safety. In: Safety and Reliability: Methodology and Applications - Proceedings of the European Safety and Reliability Conference, ESREL (2014)
8. Surashev, N., Dudkin, M., Yelemes, D., Kalieva, A.: The planetary vibroexciter with elliptic inner race. Advanced Materials Research, vol. 694–697, pp 229–232. Trans Publications, Switzerland (2013). https://doi.org/10.4028/www.scientific. net/AMR.694-697.229
9. Doudkin, M.V., Fadeyev, S.N., Pichugin, S.Y.: Studying the machines for road maintenance. Life Sci. J. **10**(12), 134–138 (2013). (ISSN1097–8135). http://www.lifesciencesite.com
10. Doudkin, M.V., Pichugin, S.Y.U., Fadeyev, S.N.: The analysis of road machine working elements parameters. World Appl. Sci. J. **23**(2), 151–158 (2013). (ISSN/E-ISSN: 1818–4952/1991-6426). IDOSI Publications. https://doi.org/10.5829/idosi. wasj.2013.23.02.13061
11. Doudkin, M., Kim, A., Kim, V.: Application of finite elements method for modeling and analysis of feed elements stresses of the vibroscreen. In: Proceedings of the 14th International Scientific Conference: Computer Aided Engineering, Politechnika Wroclawska, 8 p. (2018)
12. Kapaeva, S., Bergander, M., Vakhguelt, A., Khairaliyev, S.: Ultrasonic evaluation of the combined effect of corrosion and overheating in grade 20 steel water-wall boiler tubes. INSIGHT **59**(12), 637–643 (2017). ISSN 1354–2575

13. Azamatov, B.N., Ozhikenov, K.A., Azamatova, Z.K.: Assessment of the effectiveness of the use of palladium in catalytic SHS-units for diesel engines. News Natl. Acad. Sci. Repub. Kazakhstan, Ser. Geol. Tech. Sci. **4**(424), 142–147 (2017)
14. Ozhikenov, K.A., Mikhailov, P.G., Ismagulova, R.S., Azamatova, Z.K., Azamatov, B.N.: Development of technologies, methods and devices of the functional diagnostics of microelectronic sensors parts and components. In: 2016 13th International Scientific-Technical Conference on Actual Problems of Electronic Instrument Engineering, APEIE 2016 - Proceedings, vol. 1, pp. 84–90 (2016). https://doi.org/10. 1109/APEIE.2016.7802218
15. Azamatov, B.N., Kvassov, A.I., Azamatova, Z.K.: Hydrocyclones set ACS with variable geometry in the HAR TPP system. In: 2017 International Conference on Industrial Engineering, Applications and Manufacturing, ICIEAM 2017 - Proceedings (2017). https://doi.org/10.1109/ICIEAM.2017.8076132
16. Kapaeva, S., Bergander, M., Vakhguelt, A., Khairaliyev, S.: Remaining life assessment for boiler tubes affected by combined effect of wall thinning and overheating. J. Vibroengineering JVE **19**(8), 5892–5907 (2017)
17. Kapaeva, S., Bergander, M., Vakhguelt, A.: Combination non-destructive test (NDT) method for early damage detection and condition assessment of boiler tubes. Procedia Eng. **188**, 125–132 (2017). ISSN 1877-7058

Applications of Parallel Computing Technologies for Modeling the Flow Separation Process behind the Backward Facing Step Channel with the Buoyancy Forces

A. Issakhov$^{(\boxtimes)}$, A. Abylkassymova, and M. Sakypbekova

Al-Farabi Kazakh National University, Almaty, Kazakhstan
alibek.issakhov@gmail.com

Abstract. Taking into account the high rate of construction in the modern big cities, it is very important to save the natural aerodynamics between the buildings. It is necessary to explore the ventilation of space between architectural structures, making a preliminary prediction before construction starting. The most optimal way of evaluating is to build a mathematical model of air flow. This paper presents numerical solutions of the wind flow around the architectural obstacles with the vertical buoyancy forces. An incompressible Navier-Stokes equation is used to describe this process. This system is approximated by the control volume method and solved numerically by the projection method. The Poisson equation that is satisfying the discrete continuity equation solved by the Jacobi iterative method at each time step. For check correctness of mathematical model and numerical algorithm is solved test problem. The numerical solutions of the backward-facing step flow with the vertical buoyancy forces, which was compared with the numerical results of other authors. This numerical algorithm is completely parallelized using various geometric domain decompositions (1D, 2D and 3D). Preliminary theoretical analysis of the various decomposition methods effectiveness of the computational domain and real computational experiments for this problem were made and the best domain decomposition method was determined. In the future, a proven mathematical model and parallelized numerical algorithm with the best domain decomposition method can be applied for various complex flows with the vertical buoyancy forces.

Keywords: Domain decomposition method ·
Flow around the architectural obstacles · Backward-facing step flow ·
Projection method · Vertical buoyancy forces · Mixed convection

1 Introduction

The increased pace of construction in modern large cities and, in particular, Almaty, leads to a tightening of architectural structures. Due to the increase in

© Springer Nature Switzerland AG 2019
Y. Shokin and Z. Shaimardanov (Eds.): CITech 2018, CCIS 998, pp. 97–113, 2019.
https://doi.org/10.1007/978-3-030-12203-4_10

the population of cities and to save space, mostly high-rise multi-storey buildings are being built. As a consequence, this entails such consequences as a violation of the natural aerodynamics of the city, which in turn leads to increased gas contamination of the city, the accumulation of heavy metals in the lower atmosphere, and to the violation of the local climate. The building codes and norms currently used in the construction and design of buildings do not contain aerodynamic criteria and coefficients indicating the optimal distance between buildings of different heights. When determining these standards, various natural and climatic features are taken into account, such as wind loads, insolation, etc. Fire safety requirements are also taken into account. However, the above-mentioned documents do not take into account the factor of natural aerodynamics of space between neighboring buildings. The distance between buildings and structures is considered to be the distance between the outer walls or other structures. As a result, when designing, the distances between building objects are laid, which can not provide free movement of the wind vortex, which leads to a disturbance of the natural air flow. In this thesis, a model of aerodynamics between two high-rise buildings is considered. This mathematical model allows you to accurately calculate the optimal distance between the two buildings, which will take into account the climatic features and will preserve the natural purge. In many technical flows of practical interest, like flow divisions, with the sudden expansion of geometry or with subsequent re-joining, are a common occurrence. The existence of a flow separation and recirculation area has a significant effect on the performance of heat transfer devices, for example, cooling equipment in electrical engineering, cooling channels of turbine blades, combustion chambers and many other heat exchanger surfaces that appear in the equipment. Many papers are devoted to the motion of a fluid with separation and reconnection of flows without taking into account the buoyancy forces. The importance of this process is indicative of the number of papers where special attention was paid to building equipment [1–3] and developing experimental and theoretical methods for detailed study of flows with separation regions [4–7, 28–30]. An extensive survey of isothermal flows in fluid flows is given in papers [10–12]. Heat transfer in the flows has been investigated by many authors, like Aung [13, 14], Aung et al. [15], Aung and Worku [16], Sparrow et al. [17, 18] and Sparrow and Chuck [19]. However, published papers on this topic do not take into account the strength of buoyancy force on the flow stream or the characteristics of heat transfer. These effects become significant in the laminar flow regime, where the velocity is relatively low, and when the temperature difference is relatively high. Ngo and Byon [26] studied the location effect of the heater and the size of the heater in a two-dimensional square cavity using the finite element method. Oztop and Abu-Nada [27] numerically investigated natural convection in rectangular shells, partially heated from the side wall by the finite volume method. In this paper considered the influence of buoyancy forces on the flow and heat transfer characteristics in individual flows. Numerical solutions for a laminar mixed convective airflow ($Pr = 0.7$) in a vertical two-dimensional channel with a backward-facing step to maintain the buoyancy effect are shown in Fig. 1. Numerical results of interest, such as velocity and temperature distributions, re-binding lengths and friction coefficients are presented for the purpose of illustrating the effect of buoyancy forces on these parameters.

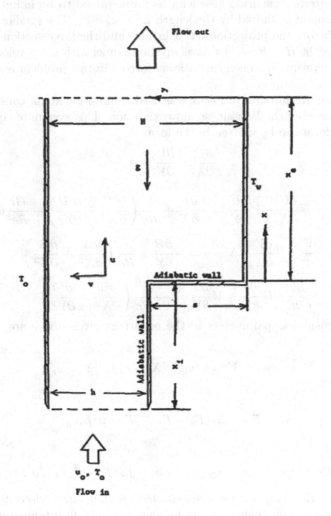

Fig. 1. Schematic representation of the backward-facing step flows.

2 Mathematical Formulation of the Problem

Consider a two-dimensional laminar convective flow in a vertical channel with a sudden expansion behind the inverse step of height s, as shown in Fig. 1. The straight wall of the channel is maintained at a uniform temperature equal to the temperature of the inlet air T_0. The stepped wall below the stage is heated to a uniform temperature, which can be adjusted to any desired value T_w. The upper part of the stepped wall and the reverse side is installed as an adiabatic surface. The inlet length of the channel x_i and the outlet lower length x_e of the channel are appropriate dimensions. These lengths are assumed to be infinite, but the simulation domain is limited by the length $L_e = x_e + x_i$. The smaller section of the channel before the projection has a height, and the large section below the stage has a height $H = h + s$. Air flows up the channel with mean velocity u_0 and uniform temperature T_0. The gravitational force g in this problem is considered to act vertically downwards.

To describe this physical problem, was used assumption about constant properties, and was used the Boussinesq approximation. This system of equations in an immense form can be written in the form:

$$\frac{\partial U}{\partial X} + \frac{\partial V}{\partial Y} = 0 \tag{1}$$

$$\frac{\partial U}{\partial t} + U\frac{\partial U}{\partial X} + V\frac{\partial U}{\partial Y} = -\frac{\partial P}{\partial X} + \frac{1}{Re}\left(\frac{\partial^2 U}{\partial X^2} + \frac{\partial^2 U}{\partial Y^2}\right) + \frac{Gr}{Re^2}\theta \tag{2}$$

$$\frac{\partial V}{\partial t} + U\frac{\partial V}{\partial X} + V\frac{\partial V}{\partial Y} = -\frac{\partial P}{\partial Y} + \frac{1}{Re}\left(\frac{\partial^2 V}{\partial X^2} + \frac{\partial^2 V}{\partial Y^2}\right) \tag{3}$$

$$\frac{\partial \theta}{\partial t} + U\frac{\partial \theta}{\partial X} + V\frac{\partial \theta}{\partial Y} = \frac{1}{Pr\, Re}\left(\frac{\partial^2 \theta}{\partial X^2} + \frac{\partial^2 \theta}{\partial Y^2}\right) \tag{4}$$

The dimensionless parameters in the equations given above are defined by the formula:

$$U = u/u_0, \quad V = v/u_0, \quad X = x/s, \quad Y = y/s,$$

$$\theta = (T - T_0)/(T_w - T_0), \quad P = p/\rho_0 u_0^2,$$

$$Pr = \nu/\alpha, \quad Re = u_0 s/\nu, \quad Gr = g\beta(T_w - T_0)s^3/\nu^2.$$

Where α – the temperature diffusion, ν – the kinematic viscosity, and β – the thermal expansion coefficient are estimated at the film temperature $T_f = (T_0 + T_w)/2$ (Fig. 2).

Boundary conditions:

(a) Inlet conditions: At the point $X = -X_i$ and $1 \leq Y \leq H/s$: $U = u_i/u_0$, $V = 0$, $\theta = 0$, $\frac{\partial p}{\partial x} = -\frac{Gr}{Re^2}\theta$.
where u_i is the local distribution of velocities at the inlet, which is assumed to have a parabolic profile and u_i/u_0 an average inlet velocity, that is, given by formula

$$u_i/u_0 = 6\left[-y^2 + (H + s)y - Hs\right]/(H - s)^2$$

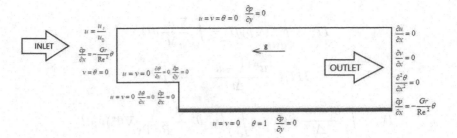

Fig. 2. Boundary conditions.

(b) Outlet conditions: At the point $X = X_e$ and $0 \leq Y \leq H/s$: $\partial U / \partial X = 0$, $\partial^2 \theta / \partial X^2 = 0$, $\partial V / \partial X = 0$, $\frac{\partial p}{\partial x} = -\frac{Gr}{Re^2} \theta$.

(c) on the top wall: At the point $Y = H/s$ and $-X_i \leq X \leq X_e$: $U = 0$, $V = 0$, $\theta = 0$, $\frac{\partial p}{\partial y} = 0$.

(d) on the wall of the upper stage: At the point $Y = 1$ and $-X_i \leq X < 0$: $U = 0$, $V = 0$, $\partial \theta / \partial Y = 0$, $\frac{\partial p}{\partial y} = 0$.

(e) on the wall of the lower stage: At point $X = 0$ and $0 \leq Y \leq 1$: $U = 0$, $V = 0$, $\partial \theta / \partial X = 0$, $\frac{\partial p}{\partial x} = 0$.

(f) on the wall below the stage: At the point $Y = 0$ and $0 \leq X \leq X_e$: $U = 0$, $V = 0$, $\theta = 1$, $\frac{\partial p}{\partial y} = 0$.

The last term on the right-hand side of Eq. (2) is the contribution of the buoyancy force. The length of the downstream flow from the simulation area was chosen to be 70 steps ($X_e = 70$). The upper length of the design area was chosen to be 5 steps (i.e. $X_i = 5$), and the velocity profile at the input area was set as parabolic profile, like $u_i / u_0 = 6[-y^2 + (H + s)y - Hs]/(H - s)^2$, and temperature was chosen as uniform T_0.

3   The Numerical Algorithm

For a numerical solution of this system of equations, the projection method is used [8,20–23,36]. The equations are approximated by the finite volume method [20,24,35–38]. At the first stage it is assumed that the transfer of momentum is carried out only through convection and diffusion, and an intermediate velocity field is calculated by the fourth-order Runge-Kutta method [21,22,31–34,36]. At the second stage, according to the found intermediate velocity field, there is a pressure field. The Poisson equation for the pressure field is solved by the Jacobi method. At the third stage it is assumed that the transfer is carried out only due to the pressure gradient. At the fourth stage, the equations for the temperature are calculated by the fourth-order Runge-Kutta method [21,22,31–34,36].

$$I. \quad \int_\Omega \frac{u^* - u^n}{\Delta t} d\Omega = -\oint_{\partial \Omega} \left(u^n \, u^* - \frac{1}{Re} \nabla u^* \right) n_i d\Gamma - \int_\Omega \frac{Gr}{Re^2} \theta d\Omega,$$

$$II. \quad \oint_{\partial\Omega} (\nabla p)d\Gamma = \int_{\Omega} \frac{\nabla \boldsymbol{u}^*}{\Delta t}d\Omega,$$

$$III. \quad \frac{\boldsymbol{u}^{n+1} - \boldsymbol{u}^*}{\Delta t} = -\nabla p,$$

$$IV. \quad \int_{\Omega} \frac{\theta^* - \theta^n}{\Delta t}d\Omega = -\oint_{\partial\Omega} (\boldsymbol{u}^n \theta^* - \frac{1}{Re \, Pr}\nabla\theta^*)n_i d\Gamma,$$

4   Parallelization Algorithm

For numerical simulation was constructed a computational mesh by using the PointWise software. The problem was launched on the ITFS-MKM software using a high-performance computing. The equations are approximated by the finite volume method (FVM) and used collocated grid, because it makes parallisation of numerical algorithm simple and efficient to use domain decomposition method. For pressure velocity coupling is used Rhie-Chow interpolation. This coupling interpolate pressure on the faces of the cell and then use this face pressure to construct the central difference scheme for pressure, which couples the adjacent pressures and avoid the checkerboard effect. For convective flux term used first order upstream flow scheme. This numerical algorithm is completely parallelized using various geometric domain decompositions (1D, 2D and 3D). Geometric partitioning of the computational grid is chosen as the main approach of parallelization. In this case, there are three different ways of exchanging the values of the grid function on the computational nodes of a one-dimensional, two-dimensional, and three-dimensional mesh. After the domain decomposition stage, when parallel algorithms are built on separate blocks, a transition is made to the relationships between the blocks, the simulations on which will be executed in parallel on each processor. For this purpose, a numerical solution of the equation system was used for an explicit scheme, since this scheme is very efficiently parallelized. In order to use the domain decomposition method as a parallelization method, this algorithm uses the boundary nodes of each subdomain in which it is necessary to know the value of the grid function that borders on the neighboring elements of the processor. To achieve this goal, at each compute node, ghost points store values from neighboring computational nodes, and organize the transfer of these boundary values necessary to ensure homogeneity of calculations for explicit formulas [36].

Data transmission is performed using the procedures of the MPI library [25]. By doing preliminary theoretical analysis of the effectiveness of various domain decomposition methods of the computational domain for this problem, which will estimate the time of the parallel program as the time T_{calc} of the sequential program divided by the number of processors plus the transmission time $T_p = T_{calc}/p + T_{com}$. While transmissions for various domain decomposition methods can be approximately expressed through capacity [36]:

$$T^{1D}_{com} = t_{send}2N^2 \times 2$$
$$T^{2D}_{com} = t_{send}2N^2 \times 4p^{1/2} \qquad (5)$$
$$T^{3D}_{com} = t_{send}2N^2 \times 6p^{2/3}$$

where N^3 – the number of nodes in the computational mesh, p - the number of processors (cores), t_{send} – the time of sending one element (number).

It should be noted that for different decomposition methods, the data transmission cost can be represented as $T^{1D}_{com} = t_{send}2N^2xk(p)$ in accordance with the formula (5), where $k(p)$ is the proportionality coefficient, which depends on the domain decomposition method and the number of processing elements used [36].

At the first stage, one common program was used, the size of the array from start to run did not change, and each element of the processor was numbered by an array of elements, starting from zero. For the test simulation is used well known problem – 3D cavity flow. Despite the fact that according to the theoretical analysis of 3D decomposition is the best option for parallelization (Fig. 3), computational experiments showed that the best results were achieved using 2D decomposition, when the number of processes varies from 25 to 144 (Fig. 3) [36].

Based on the preliminary theoretical analysis of the graphs, the following character can be noted. The simulation time without the interprocessor communications cost with different domain decomposition methods should be approximately the same for the same number of processors and be reduced by T_{calc}/p. In fact, the calculated data show that when using 2D decomposition on different computational grids, the minimal cost for simulation and the cost graphs are much higher, depending on the simulation time, on several processors taken T_{calc}/p [36].

To explain these results, it is necessary to pay attention to the assumptions made in the preliminary theoretical analysis of efficiency for this task. First, it was assumed that regardless of the distribution of data per processor element, the same amount of computational load was done, which should lead to the same time expenditure. Secondly, it was assumed that the time spent on interprocessor sending's of any degree of the same amount of data is not dependent on their memory choices. In order to understand what is really happening, the following sets of computational simulations test were carried out. For evaluation, the sequence of the first approach was considered when the program is run in a single-processor version, and thus simulates various geometric domain decomposition methods of data for the same amount of computation performed by each processor [36].

5   Numerical Results for Test Problem

Geometric parameters are indicated in Fig. 1: channel length $L = 75$, channel height $H = 2$, step height $S = 1$. Numerical results were obtained for the dimensionless numbers $Re = 50$, $Pr = 0.7$ and $Gr = 19.1$ [9].

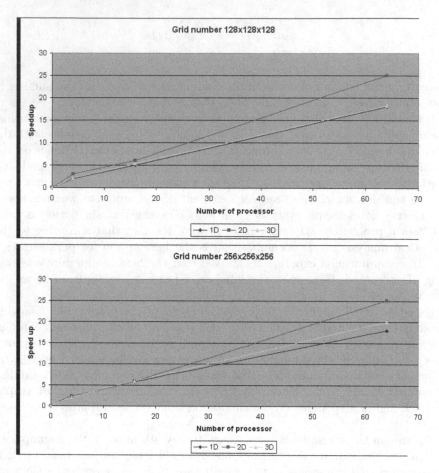

Fig. 3. Speed-up for various domain decomposition methods of the computational domain.

Figure 4 shows the comparison of the longitudinal velocity profile with the numerical data of Lin et al. [9] at the point $x/x_f = 0.5$, where $x_f = 2.91$. Figure 5 shows the comparison of temperature profiles with the numerical data of Lin et al. [9] at the point $x/x_f = 0.5$, where $x_f = 2.91$. It can be seen from the figures that the mathematical model and the numerical algorithm which is used in this paper is coincided with the numerical results obtained by Lin et al. [9]. Figure 6 shows the streamlines and the horizontal velocity contour for dimensionless numbers $Re = 50$, $Pr = 0.7$ and $Gr = 19.1$. Figure 7 shows the vertical velocity contour for dimensionless numbers $Re = 50$, $Pr = 0.7$ and $Gr = 19.1$. Figure 8 shows the temperature profile for dimensionless numbers $Re = 50$, $Pr = 0.7$ and $Gr = 19.1$. For a better understanding of this process from Figs. 6, 7 and 8 can be seen

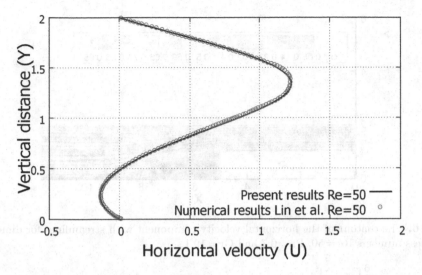

Fig. 4. Velocity profile with vertical buoyancy forces for dimensionless number Re = 50, $\Delta T = 1\,^\circ\mathrm{C}$, $x/x_f = 0.5$, where $x_f = 2.91$.

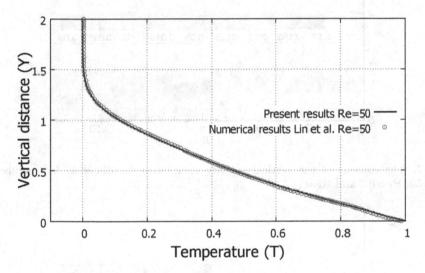

Fig. 5. Temperature profile with vertical buoyancy forces for dimensionless number Re = 50, $\Delta T = 1\,^\circ\mathrm{C}$, $x/x_f = 0.5$, where $x_f = 2.91$.

the development of the backward-facing step flow with vertical buoyancy force: the initiation and process of the development of the region of flows reconnection with taking into account the buoyancy forces.

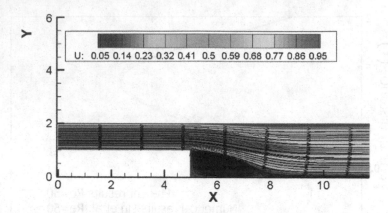

Fig. 6. The contour of the horizontal velocity component with streamlines for dimensionless numbers Re = 50, Pr = 0.7 and Gr = 19.1.

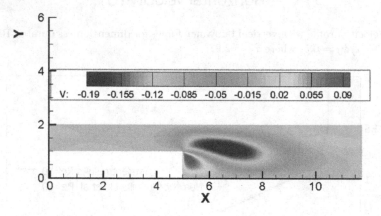

Fig. 7. The contour of the vertical velocity component for dimensionless numbers Re = 50, Pr = 0.7 and Gr = 19.1.

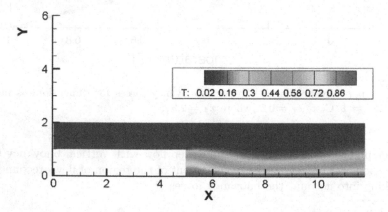

Fig. 8. Temperature contour for dimensionless numbers Re = 50, Pr = 0.7 and Gr = 19.1.

6   Numerical Results for Real Problem

For the real problem considered a man-made obstacle 9 floors (27 m) and 5 floors (15 m) buildings. The wind flow is conventionally moving from the high building side to the low one. The following models consider the calm, according to the Beaufort wind speed scale. The speed of wind is in the range from 0 to 0.2 m/s. To find the optimal distance, various parameters prescribed in the above-mentioned standards were used.

According to the fire protection requirements specified in Building norms and regulations of the Republic of Kazakhstan 3.01-01-2002, 2.12 * [18], the minimum distance between houses with a height of 4 floors or more must be at least 20 m. However, having constructed this model, a result was obtained, showing that at such a distance between the buildings, there was no wind, therefore, air circulation does not occur in this interval.

After that, the IBC (International Building Code) standard used in the USA was considered, where the distance between two buildings is calculated according to the following formula (Fig. 9) [19]:

$$\delta_{MT} = \sqrt{(\delta_{M1})^2 + (\delta_{M2})^2}$$

where δ_{MT} - required distance, δ_{M1}, δ_{M2} - height of the first and second buildings, respectively.

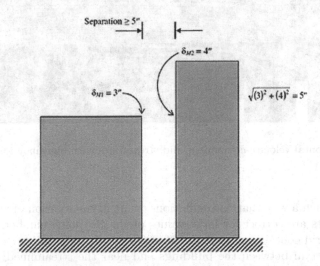

Fig. 9. The calculation of the distance between two high-rise buildings according to IBC 2009 1613.6.7

For the calculations, the area was divided into the 14 design subareas of different sizes. Each subregion is a grid block, which contains a part of a curvilinear, uneven, unstructured grid. When creating a grid, the number of points

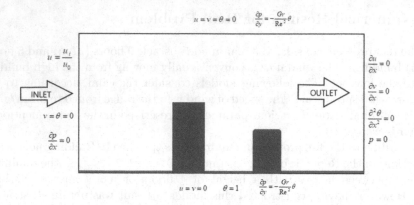

Fig. 10. Boundary conditions

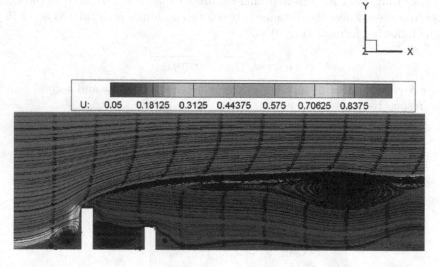

Fig. 11. Horizontal velocity-component and streamlines with buoyancy force for Pr = 0.7 and $\Delta T = 1\,^{\circ}\mathrm{C}$.

is chosen in such a way that no oscillations occur in the solution of the problem and the results are correct for large values of the Reynolds number. Thus, the constructed grid contains more than 100 000 control volumes.

In the interval between the buildings and near the streamlined surfaces of buildings, the grid is thickened, i.e. the sizes of control volumes decrease. This allows for more accurate simulations. As the grid is removed from the vortex zone and the dimensions of the control volumes increase the flow around. Therefore, in areas that are not of great interest in solving a given problem, net catch is smaller. And in the most important zones for the model, the number of nodes is greater, which allows obtaining more accurate results.

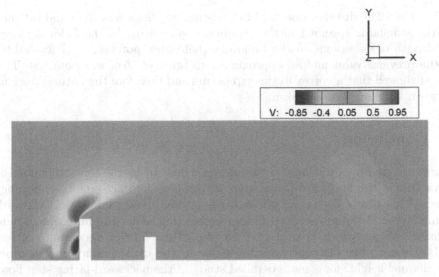

Fig. 12. Vertical velocity-component with buoyancy force for $Pr = 0.7$ and $\Delta T = 1\,^{\circ}$C.

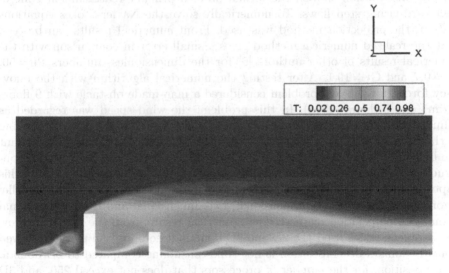

Fig. 13. Temperature component with buoyancy force for $Pr = 0.7$ and and $\Delta T = 1\,^{\circ}$C.

Boundary conditions are specified for physical variables: pressure, temperature and velocity components: p pressure, u, v, w - velocity components, θ - temperature (Fig. 10).

In Fig. 11 shows horizontal velocity-component and streamlines with buoyancy force for $Pr = 0.7$ and $\Delta T = 1\,^{\circ}$C. From Fig. 12 can be seen vertical velocity-component with buoyancy force for $Pr = 0.7$ and $\Delta T = 1\,^{\circ}$C. In Fig. 13 can be seen temperature component with buoyancy force for $Pr = 0.7$ and

$\Delta T = 1\,^\circ\text{C}$. The distance obtained for existing buildings was $35\,\text{m}$ and satisfied all the standards specified in the republican standards. In the following case, the length of extensions to the buildings (balconies, porches, etc.) was added to the previous value and an approximate distance of $35\,\text{m}$ was obtained. This model showed that a vortex in the gap occurs and therefore the natural purging between buildings is not broken.

7   Conclusion

Numerical studies of the laminar flow were carried out by the zone of joining the flows behind the backward-facing step with taking into account the buoyancy forces. This gave a deeper insight into the internal flow behind the backward-facing step and the processes of flows reconnection under the influence of temperature effects, which in turn gave an idea of the further appearance of secondary zones. The distance from the ledge to the canal boundary is 4 times the channel height, for a more detailed study of the backward-facing step flows with taking into account the buoyancy forces [9]. The numerical data of the velocity distribution showed the formation of a primary reattachment zone of backward-facing step flows. To numerically solve the Navier-Stokes equations system, the projection method was used. From numerical results can be seen that the realized numerical method gives a small error in comparison with the numerical results of other authors [9] for the dimensionless numbers $\text{Re} = 50$, $\text{Pr} = 0.7$ and $\text{Gr} = 19.1$. After testing the numerical algorithm with the buoyancy forces, for the real problem considered a man-made obstacle with 9 floors ($27\,\text{m}$) and 5 floors ($15\,\text{m}$). In this problem, the wind speed was regarded as calm, according to the Beaufort wind speed scale (from 0 to $0.2\,\text{m/s}$). According to the data obtained as a result of the numerical simulation taking into account the buoyancy forces, it can be said that the current standards and rules for construction do not guarantee the required aerodynamics of the terrain. Also in this paper is used a parallel algorithm to obtain fast numerical results. This parallel algorithm is based on one-dimensional, two-dimensional and three-dimensional domain decomposition method. The numerical results from the 3D cavity flow test problem, which used 1D, 2D and 3D domain decomposition method showed that 3D domain decomposition is not time-consuming compared to 2D domain decomposition, for the number of processors that does not exceed 250, and 3D domain decomposition has more time-consuming software implementation and the use of 2D domain decomposition is sufficient for the scope of the problem. That's why for backward-facing step flow with vertical buoyancy force is used 2D domain decomposition. It should also be noted that setting the boundary conditions is an important process. In the future, this mathematical model and a parallel numerical algorithm can be applied to various complex flows taking into account the buoyancy forces.

References

1. Abbott, D.E., Kline, S.J.: Experimental investigations of subsonic turbulent flow over single and double backward-facing steps. J. Basic Eng. **84**, 317 (1962)
2. Sebanr, A.: Heat transfer to the turbulent separated flows of air downstream of a step in the surface of a plate. J. Heat Transf. **86**, 259 (1964)
3. Goldsteinr, J., Eriksenv, L., Olsonr, M., Eckerte, R.G.: Laminar separation, reattachment and transition of flow over a downstream-facing step. J. Basic Eng. **92**, 732 (1970)
4. Durst, F., Whitelawj, H.: Aerodynamic properties of separated gas flows: existing measurements techniques and new optical geometry for the laser-Doppler anemometer. Prog. Heat Mass Transf. **4**, 311 (1971)
5. Gosmana, D., Punw, M.: Lecture notes for course entitled: calculation of recirculating flow. Heat Transf. Rep. **74**, 2 (1974)
6. Kumara, Y.S.: Internal separated flows at large Reynolds number. J. Fluid Mech. **97**, 27 (1980)
7. Chiang, T.P., Tony, W.H., Sheu, Fang C.C.: Numerical investigation of vortical evolution in backward-facing step expansion flow. Appl. Math. **23**, 915–932 (1999)
8. Fletcher, C.A.J.: Computational Techniques for Fluid Dynamics, vol. 1, p. 387. Springer, New York (1988). https://doi.org/10.1007/978-3-642-58229-5
9. Lin, J.T., Armaly, B.F., Chen, T.S.: Mixed convection in buoyancy-assisting, vertical backward-facing step flows. Int. J. Heat Mass Transf. **33**(10), 2121–2132 (1990)
10. Armaly, B.F., Durst, F.: Reattachment length and recirculation regions downstream of two dimensional single backward facing step. In: Momentum and Hear Transfer Process in Recirculating Flows, ASME HTD, vol. 13, pp. 1–7. ASME, New York (1980)
11. Eaton, J.K., Johnson, J.P.: A review of research on subsonic turbulent flow reattachment. AIAA J. **19**, 1093–1100 (1981)
12. Simpson, R.L.: A review of some phenomena in turbulent flow separation. J. Fluid Eng. **103**, 520–530 (1981)
13. Aung, W.: An experimental study of laminar heat transfer downstream of backsteps. J. Heat Transf. **105**, 823–829 (1983)
14. Aung, W.: Separated forced convection. In: Proceedings of ASMEIJSME Thermal Enana Joint Conference, vol. 2, pp. 499–515. ASME, New York (1983)
15. Aung, W., Baron, A., Tsou, F.K.: Wall independency and effect of initial shear-layer thickness in separated flow and heat transfer. Int. J. Hear Muss Transf. **28**, 1757–1771 (1985)
16. Aung, W., Worku, G.: Theory of fully developed, combined convection including flow reversal. J. Hear Transf. **108**, 485–488 (1986)
17. Sparrow, E.M., Chrysler, G.M., Azevedo, L.F.: Observed flow reversals and measured-predicted Nusselt numbers for natural convection in a one-sided heated vertical channel. J. Heat Transf. **106**, 325–332 (1984)
18. Sparrow, E.M., Kang, S.S., Chuck, W.: Relation between the points of flow reattachment and maximum heat transfer for regions of flow separation. Int. J. Hear Mass Transf. **30**, 1237–1246 (1987)
19. Sparrow, E.M., Chuck, W.: PC solutions for heat transfer and fluid flow downstream of an abrupt, asymmetric enlargement in a channel. Numer. Hear Transf. **12**, 1940 (1987)
20. Chung, T.J.: Computational Fluid Dynamics (2002)

21. Issakhov, A.: Mathematical modeling of the discharged heat water effect on the aquatic environment from thermal power plant. Int. J. Nonlinear Sci. Numer. Simul. **16**(5), 229–238 (2015)
22. Issakhov, A.: Mathematical modeling of the discharged heat water effect on the aquatic environment from thermal power plant under various operational capacities. Appl. Math. Model. **40**(2), 1082–1096 (2016)
23. Issakhov, A.: Large eddy simulation of turbulent mixing by using 3D decomposition method. J. Phys.: Conf. Ser. **318**(4), 1282–1288 (2011)
24. Chorin, A.J.: Numerical solution of the Navier-Stokes equations. Math. Comp. **22**, 745–762 (1968)
25. Karniadakis, G.E., Kirby II, R.M.: Parallel Scientific Computing in C++ and MPI: A Seamless Approach to Parallel Algorithms and their Implementation. Cambridge University Press, Cambridge (2000)
26. Ngo, I., Byon, C.: Effects of heater location and heater size on the natural convection heat transfer in a square cavity using finite element method. J. Mech. Sci. Technol. **29**(7), 2995 (2015)
27. Oztop, H.F., Abu-Nada, E.: Numerical study of natural convection in partially heated rectangular enclosures filled with nanofluids. Int. J. Heat. Fluid Flow **29**(5), 1326–1336 (2008)
28. Xie, W.A., Xi, G.N.: Geometry effect on flow fluctuation and heat transfer in unsteady forced convection over backward and forward facing steps. Energy **132**, 49–56 (2017)
29. Yilmaz, I., Oztop, H.F.: Turbulence forced convection heat transfer over double forward facing step flow. Int. Commun. Heat Mass Transf. **33**, 508–517 (2006)
30. Xie, W.A., Xi, G.N., Zhong, M.B.: Effect of the vortical structure on heat transfer in the transitional flow over a backward-facing step. Int. J. Refrig. **74**, 463–472 (2017)
31. Issakhov, A.: Modeling of synthetic turbulence generation in boundary layer by using zonal RANS/LES method. Int. J. Nonlinear Sci. Numer. Simul. **15**(2), 115–120 (2014)
32. Issakhov, A.: Numerical modelling of distribution the discharged heat water from thermal power plant on the aquatic environment. In: AIP Conference Proceedings, vol. 1738, p. 480025 (2016)
33. Issakhov, A.: Numerical study of the discharged heat water effect on the aquatic environment from thermal power plant by using two water discharged pipes. Int. J. Nonlinear Sci. Numer. Simul. **18**(6), 469–483 (2017)
34. Issakhov, A.: Numerical modelling of the thermal effects on the aquatic environment from the thermal power plant by using two water discharge pipes. In: AIP Conference Proceedings, vol. 1863, p. 560050 (2017)
35. Issakhov, A., Zhandaulet, Y., Nogaeva, A.: Numerical simulation of dam break flow for various forms of the obstacle by VOF method. Int. J. Multiph. Flow **109**, 191–206 (2018)
36. Issakhov, A., Abylkassymova, A.: Application of parallel computing technologies for numerical simulation of air transport in the human nasal cavity. In: Zelinka, I., Vasant, P., Duy, V.H., Dao, T.T. (eds.) Innovative Computing, Optimization and Its Applications. SCI, vol. 741, pp. 131–149. Springer, Cham (2018). https://doi.org/10.1007/978-3-319-66984-7_8

37. Issakhov, A., Mashenkova, A.: Numerical study for the assessment of pollutant dispersion from a thermal power plant under the different temperature regimes. Int. J. Environ. Sci. Technol. 1–24 (2019). https://doi.org/10.1007/s13762-019-02211-y
38. Issakhov, A., Bulgakov, R., Zhandaulet, Y.: Numerical simulation of the dynamics of particle motion with different sizes. Eng. Appl. Comput. Fluid Mech. **13**(1), 1–25 (2019)

Numerical Study of a Passive Scalar Transport from Thermal Power Plants to Air Environment

A. Issakhov[✉] and A. Baitureyeva

Al-Farabi Kazakh National University,
71, al-Farabi Avenue, 050040 Almaty, Republic of Kazakhstan
alibek.issakhov@gmail.com, abatur@yandex.kz

Abstract. This paper presents computational fluid dynamics (CFD) techniques in modeling the pollution distribution from thermal power plant. Carbon dioxide (CO_2) dispersion from a thermal power plant was simulated. The mathematical model and numerical algorithm were tested using a test problem and gave a good match with the experimental data. The influence of gravity force was taken into account. The k-epsilon model of turbulence with the buoyancy force was used. Calculations were performed by ANSYS Fluent. As a result, there was found a distance from the source at which the impurity settles on the ground surface.

Keywords: Navier-Stokes equations · Mass transfer ·
Numerical simulation · Air pollution · Concentration

1 Introduction

Energy is the most significant among the industries that have a negative impact on the environment. This is due to the fact that the development of society and the population growth constantly require more and more energy. As a consequence, emissions of pollutants into the atmosphere should also increase. Therefore, studying the nature of these emissions, their structure, the impact of these pollutants on the elements of the environment - is one of the urgent tasks of modern applied ecology. The structure of pollution emissions into the atmosphere depends on the capacity of emission sources, the location of energy facilities in relation to ecologically significant areas, and the physical nature of emissions. For example, discharges from the coal industry and from TPPs are close in magnitude, but they substantially exceed the discharges from the gas industry. NPPs shed much less sulfates, chlorides and nitrates than TPPs, but more phosphorus emits out of nuclear power plants. In addition, it should be noted that mercury, selenium, fluorine and other elements, that are not completely captured by the waste gas filtration system, become a source of air pollution in coal combustion products at high-capacity power plants (over 2 million kW). Volatile (lead, copper, zinc, cerbium, itterium, etc.) elements are distributed between solid combustion products, which requires special procedures for the disposal of ash and

© Springer Nature Switzerland AG 2019
Y. Shokin and Z. Shaimardanov (Eds.): CITech 2018, CCIS 998, pp. 114–124, 2019.
https://doi.org/10.1007/978-3-030-12203-4_11

slag wastes. In addition, the nature of the impact on ecological systems depends on the natural conditions of the location area and the physical nature of the ejected ingredient. Recently, much attention has been paid to the effects of various pollutants on environmental and human elements, as well as on models and physical principles of the spread of these pollutants in environmental objects.

At the same time, the issues of the emissions spread and their impact on ecological systems have not been adequately worked out. The most successful is the assessment of the effect of thermal emissions and discharges on the thermal regime of rivers and reservoirs. However, the study of thermal emissions and the associated effects of changing microclimatic conditions, impact on terrestrial ecological systems, require further elaboration taking into account the concepts of sustainable ecological development of ecosystems, monitoring systems and environmental safety. The purpose of this work is to assess the impact of large-scale energy emissions on the environment based on a numerical model of emissions diffusion from sources. One of the pipes of Ekibastuz SDPP-1 (Kazakhstan) was chosen as the real object of the research. Its height of stack is 330 m and the diameter is 10 m.

2   Mathematical Model

A detailed description of the latest works about the study of the jet distribution in the crossflow is given in [1] and [2]. The velocity field is numerically calculated in papers [3–7]. The passive scalar mass fraction field was considered in papers [8–10]. CFD is often used to solve these problems. Reynolds-averaged Navier-Stokes equations (RANS) were used and the results were compared with the experimental data in papres [11–16]. Good correspondence between numerical solutions (DNS) and experiments was obtained by [17] and [18]. However, the DNS simulation requires large computational costs, which is unacceptable for solving large-scale problems in real scales [1] and [19–23]. As a result, the k-epsilon turbulence model was used in this paper. The CFD simulations of such processes are based on the resolution of the Navier-Stokes equations (continuity of mass and momentum equations) [24–28]. Since the RANS equation was applied taking into account the buoyancy, the equations for turbulent kinetic energy and dissipation are presented as follows:

$$\frac{\partial k}{\partial t} + \frac{\partial}{\partial x_j}(u_j k) = \frac{\partial}{\partial x_j}\left[\left(\mu + \frac{\mu_t}{\sigma_k}\right)\frac{\partial k}{\partial x_j}\right] + P_k - \rho\varepsilon + P_{kb}$$

$$\frac{\partial \varepsilon}{\partial t} + \frac{\partial}{\partial x_j}(u_j \varepsilon) = \frac{\partial}{\partial x_j}\left[\left(\mu + \frac{\mu_t}{\sigma_\varepsilon}\right)\frac{\partial \varepsilon}{\partial x_j}\right] + \frac{\varepsilon}{k}(C_{\varepsilon 1}P_k - C_{\varepsilon 2}\rho\varepsilon + C_{\varepsilon 1}P_{\varepsilon b})$$

The term P_k represents the generation of turbulence kinetic energy due to the mean velocity gradients. P_{kb}, $P_{\varepsilon b}$ represent the buoyancy forces, where $P_{kb} = -\frac{\mu_t}{\rho\sigma_\rho}\rho\beta g_i\frac{\partial T}{\partial x_i}$ and $P_{\varepsilon b} = C_3 max\,(0, P_{kb})$. Here β is thermal expansion coefficient, $\sigma_\rho = 0.9, C_{s1}, C_{s2}, \sigma_k, \sigma_\varepsilon$ are constants. For solving conservation equations for chemical species, ANSYS Fluent predicts the local mass fraction of each species Y_i, through the solution of a convection-diffusion equation for the i-th species.

$$\tfrac{\partial}{\partial t}\left(\rho Y_i\right) + \nabla\left(\rho \boldsymbol{u} Y_i\right) = -\nabla \boldsymbol{J}_i + R_i + S_i$$

Here R_i is the net rate of production of species i by chemical reaction and S_i is the rate of creation by addition from the dispersed phase plus any user-defined sources.

For the turbulent flows, the mass diffusion is computed in the following form:

$$\boldsymbol{J}_i = -\left(\rho D_{i,m} + \tfrac{\mu_t}{Sc_t}\right)\nabla Y_i$$

Where $Sc_t = \tfrac{\mu_t}{\rho D_t}$ – turbulent Schmidt number (μ_t - turbulent viscosity and D_t - turbulent diffusivity).

3 Test Problem

An experimental study for this test problem was conducted in a low speed wind tunnel on a row of six rectangular jets injected at 90^0 to the crossflow [1,29] and [30]. Mean velocities were measured using a three-component laser Doppler velocimeter operating in coincidence-mode. Seeding of both jet and cross stream air was achieved with a commercially available smoke generator. To complement the detailed measurements, flow visualization was accomplished by transmitting the laser beam through a cylindrical lens, thereby generating a narrow, intense sheet of light.

3.1 Computational Domain and Grid

The computational domain included the jet channel and the space above the flat plate. A square jet was used in this study and jet diameter (jet width) was D, which was used as the length unit. The origin of the coordinate system was located at the center of the jet exit. The calculation area of the test problem is a three-dimensional channel with the pipe entering into it. The height of the crossflow channel is 20D, jet channel length is 5D, the length of the crossflow channel is 45D, the width is 3D, and the center of the pipe located at 5D from the inlet (See Fig. 1). Number of grid points: main channel $230 \times 100 \times 21$ and jet channel $7 \times 30 \times 7$ nodes. Grid was unstructured, total number of nodes was 533 697 [1].

3.2 Boundary Conditions and Flow Characteristics

The ratio of the jet velocity to the crossflow velocity is denoted by R, and is expressed as:

$$R = V_{jet}/V_{crossflow}$$

In papers [1] and [30] considered various R (0.5, 1.0 and 1.5). In present paper R = 0.5 was considered, that is why the jet velocity was 5.5 m/s, crossflow velocity was 11 m/s. Water was chosen as the material of the crossflow and jet

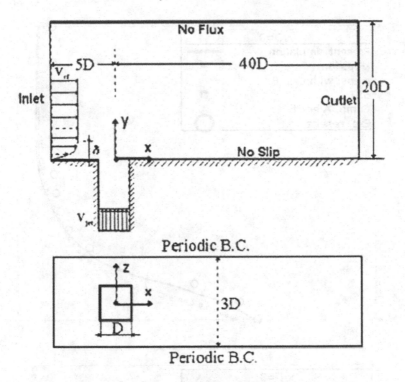

Fig. 1. Configuration of test problem computational domain.

fluid. The jet diameter was $D = 12.7$ mm. Based on the above data, the Reynolds number is defined as:

$$Re_{jet} = \rho V_{jet} D / \mu = 4700$$

Five types of boundary conditions were used: inlet, outlet, no flux, wall, periodic (See Fig. 1). According to the experimental data, the boundary layer thickness is 2D. To describe the initial crossflow velocity profile in the boundary layer, 1/7 power law wind profile was used:

$$\frac{u}{u_r} = \left(\frac{z}{z_r}\right)^{\alpha}$$

Here u is the wind velocity at height z and u_r is the known wind velocity at a reference height z_r. α is empirically obtained coefficient, which varies depending on the atmosphere stability. Here, $\alpha = 1/7$ for neutral stability conditions. Uniform velocity (11 m/s) was defined above the boundary layer 2D [1].

3.3   Comparison of Numerical Results

Figure 2 shows a comparison of the numerical results of present paper with experimental data and computational solutions of other authors (R = 0.5). The red solid line marks the results of present paper, round-shaped of experimental data

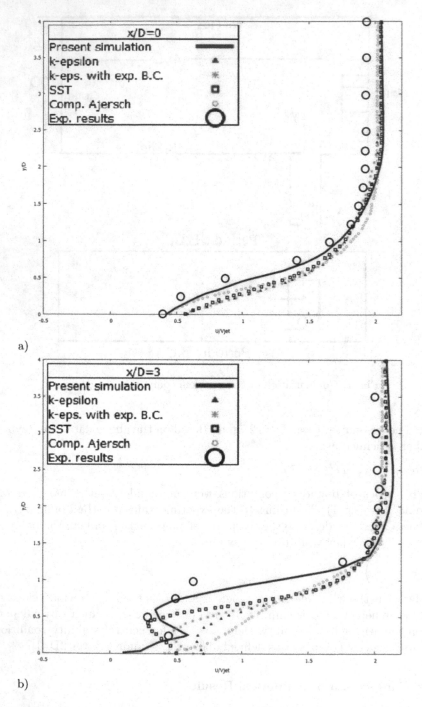

Fig. 2. Comparison of results for streamwise velocity at jet center plane (z/D = 0) for R = 0.5: (a) x/D = 0; (b) x/D = 3. (Color figure online)

(See [29]) and the rest lines illustrate computational solutions of other authors (See [30]). The results for the SST model are shown by blue squares. Comparative analysis showed that the k-epsilon model and the SIMPLE algorithm with finite volume method yield closest results to the experimental data (See Fig. 2b) [1].

The solutions obtained in this paper turned out to be more accurate than those computational solutions obtained by other authors ([30]). Values of the red line (u/Vjet) at $y/D = 0$ close to zero, when other results close to 0.5–0.7. This is more accurate from a physical point of view, since it is a near-wall field. Moreover, in the interval y/D 0.5–1.5 it is noticeable that the current calculations are much closer to the experimental data than the others. The reason is the quality of the grid: in present paper unstructured grid was used, the number of nodes was 533 697, while in paper [30] used structured grid and the number of nodes was 265 000 [1].

3.4   Geometry, Grid and Boundary Conditions

The computational domain is a three-dimensional box with a pipe inside it (See Fig. 3). The boundary conditions were set analogically to test problem: inlet, wall, outlet, symmetry, periodic. The length of the geometry was 2500 m, the height was 1500 m, the width was 200 m. The pipe was located at 250 m from the entrance of the wind.

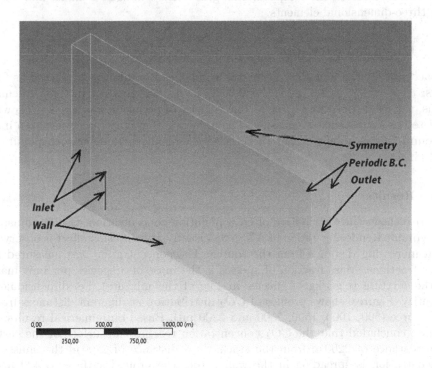

Fig. 3. Configuration of thermal power plant computational domain.

Fig. 4. Computational mesh

An unstructured grid was constructed. It was refined in the area of the pollution motion trajectory (See Fig. 4). Around the pipe, the grid size was 1 m, with 1.1 growth factor, in the remaining refined region the maximum grid size was 6 m. As a result, the computational grid consists of 822 056 nodes and 4 788 517 three-dimensional elements.

3.5 Flow Characteristics

To account for the boundary layer, the wind speed profile was described as follows: $v_x = v_{wind} \cdot (0.2371 \cdot ln\,(Y + 0.00327) + 1.3571)\,\left[\text{ms}^{-1}\right]$. In these calculations, the wind speed was 1.5 m/s. The velocity of pollution was 5 m/s. CO_2 was selected as a substance of emission. The influence of gravity force was taken into account. It was accepted that there is no chemical reaction between pollution and air.

3.6 Results

Figure 5 shows the distribution of CO_2 pollution concentration, visualized using the Volume Rendering option in ANSYS. Visually, the diffusion effect is observed with increasing distance from the source. The concentration was measured in mass fractions. Mass fraction of species is the mass of a species per unit mass of the mixture (e.g., kg of species in 1 kg of the mixture), i.e. dimensionless quantity. Figure 6 shows profiles of CO_2 distribution at different distances from the source (500, 1000, 1500, 2000 and 2200 [m]). Based on numerical results, it can be concluded that the CO_2 concentration reaches the surface of the earth at a distance of 2200 m from the stack. At a distance of 2200 m the emission concentration is spreading in the range from the ground surface to 420 m in height.

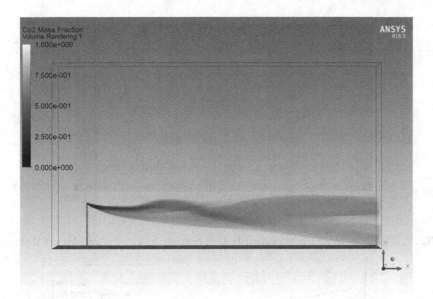

Fig. 5. The isosurface of CO_2 distribution

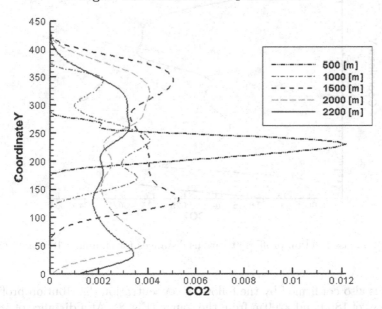

Fig. 6. CO_2 mass fraction profiles at various distances for chimney: 500, 1000, 1500, 2000, 2200 [m]

The Fig. 7 shows a CO_2 pollution distribution contour at the plane XY. Pollution settles on the ground surface at a distance of about 2220 m from the origin (i.e, 1970 m from the source).

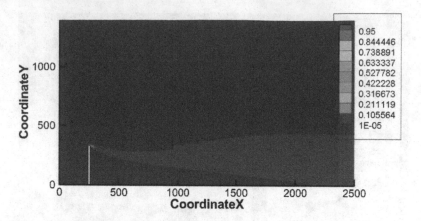

Fig. 7. The CO_2 mass fraction distribution contour at the center plane XY slice

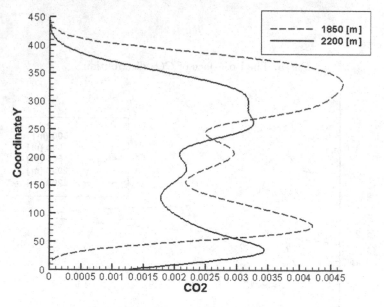

Fig. 8. CO_2 mass fraction profiles at various distances for chimney: 1850 and 2200 [m]

This is also confirmed by the following concentration distribution profiles at a distances of 1850 and 2200 m from the source (Fig. 8). At a distance of 1850 m, the concentration does not reach the ground and spreads at a height of about 20 m, while at a distance of 2200 m the concentration already reaches the ground. It can also be noted that the level of concentration decreases with increasing distance from the source: at a distance of 1850 m the maximum value of the pollution mass fraction is 0.0045, while at a distance of 2200 m the maximum value reached only 0.0035.

4   Conclusion

The purpose of this investigation was to study the dynamics of the emissions spread from energy objects. The computational fluid dynamics (CFD) techniques in modeling the pollution distribution from thermal power plant were considered. The mathematical model, turbulent model and numerical algorithm were tested using a test problem and gave a good match with the experimental data. The results obtained in this paper were closer to experimental, in comparison with the results of other authors. This was influenced by the quality of the grid: the unstructured and refined mesh gave a better result than the structured one. Carbon dioxide (CO_2) dispersion from a thermal power plant was simulated. The influence of gravity force was taken into account. To close the RANS equations was used the k-epsilon turbulent model. No additional dispersion model was applied. All simulations were performed by ANSYS Fluent. As a result, there was found a distance from the source at which the impurity settles on the ground surface. According to the obtained data, with increasing distance from the source, the concentration of pollution spreads wider under the influence of diffusion. The further the distance from the pipe then the lower the concentration. Thus, the obtained numerical data may allow in the future to predict the optimal distance from residential areas for constructing a TPP at which the emission concentrations will be at a safe level. These studies are useful for those who are interested in gas pollutant distribution in the atmosphere.

References

1. Issakhov, A., Baitureyeva, A.: Numerical modelling of a passive scalar transport from thermal power plants to air environment. Adv. Mech. Eng. **10**(10), 1–14 (2018)
2. Margason, R.J.: Fifty years of jet in crossflow research. In: AGARD Symposium on a Jet in Cross Flow, Winchester, UK. AGARD CP 534 (1993)
3. Kamotani, Y., Greber, I.: Experiments on turbulent jet in a crossflow. AIAA J. **10**, 1425–1429 (1972)
4. Fearn, R.L., Weston, R.P.: Vorticity associated with a jet in crossflow. AIAA J. **12**, 1666–1671 (1974)
5. Andreopoulos, J., Rodi, W.: Experimental investigation of jets in a crossflow. J. Fluid Mech. **138**, 93–127 (1984)
6. Krothapalli, A., Lourenco, L., Buchlin, J.M.: Separated flow upstream of a jet in a crossflow. AIAA J. **28**, 414–420 (1990)
7. Fric, T.F., Roshko, A.: Vortical structure in the wake of a transverse jet. J. Fluid Mech. **279**, 1–47 (1994)
8. Smith, S.H., Mungal, M.G.: Mixing, structure and scaling of the jet in crossflow. J. Fluid Mech. **357**, 83–122 (1998)
9. Su, L.K., Mungal, M.G.: Simultaneous measurement of scalar and velocity field evolution in turbulent crossflowing jets. J. Fluid Mech. **513**, 1–45 (2004)
10. Shan, J.W., Dimotakis, P.E.: Reynolds-number effects and anisotropy in transverse-jet mixing. J. Fluid Mech. **566**, 47–96 (2006)
11. Broadwell, J.E., Breidenthal, R.E.: Structure and mixing of a transverse jet in incompressible flow. J. Fluid Mech. **148**, 405–412 (1984)

12. Karagozian, A.R.: An analytical model for the vorticity associated with a transverse jet. AIAA J. **24**, 429–436 (1986)
13. Hasselbrink, E.F., Mungal, M.G.: Transverse jets and jet flames. Part 1. Scaling laws for strong transverse jets. J. Fluid Mech. **443**, 1–25 (2001)
14. Muppidi, S., Mahesh, K.: Study of trajectories of jets in crossflow using direct numerical simulations. J. Fluid Mech. **530**, 81–100 (2005)
15. Issakhov, A., Baitureyeva, A.: Numerical study of a passive scalar transport in lower atmosphere layer from thermal power plants. In: American Institute of Physics Conference Proceedings, vol. 1978, p. 470052 (2018). https://doi.org/10.1063/1.5044122
16. Muppidi, S., Mahesh, K.: Direct numerical simulation of passive scalar transport in transverse jets. J. Fluid Mech. **598**, 335–360 (2008)
17. Chochua, G., et al.: A computational and experimental investigation of turbulent jet and crossflow interaction. Numer. Heat Transf. A **38**, 557–572 (2000)
18. Acharya, S., Tyagi, M., Hoda, A.: Flow and heat transfer predictions for film-cooling. Ann. NY Acad. Sci. **934**, 110–125 (2001)
19. Issakhov, A.: Large eddy simulation of turbulent mixing by using 3D decomposition method. J. Phys. Conf. Ser. **318**(4), 1282–1288 (2011). https://doi.org/10.1088/1742-6596/318/4/042051
20. Issakhov, A.: Mathematical modeling of the discharged heat water effect on the aquatic environment from thermal power plant. Int. J. Nonlinear Sci. Numer. Simul. **16**(5), 229–238 (2015). https://doi.org/10.1515/ijnsns-2015-0047
21. Issakhov, A.: Mathematical modeling of the discharged heat water effect on the aquatic environment from thermal power plant under various operational capacities. Appl. Math. Model. **40**(2), 1082–1096 (2016). https://doi.org/10.1016/j.apm.2015.06.024
22. Issakhov A., Baitureyeva A.R.: Numerical simulation of a passive scalar transport from thermal power plants. In: American Institute of Physics Conference Proceedings, vol. 1836, p. 020019 (2017). https://doi.org/10.1063/1.4981959
23. Issakhov, A., Baitureyeva, A.R.: Numerical simulation of a pollutant transport in lower atmosphere layer from thermal power plants. WSEAS Trans. Fluid Mech. **12**, 33–42 (2017). Article no. n4
24. Ferziger, J.H., Peric, M.: Computational Methods for Fluid Dynamics, 3rd edn. Springer, Heidelberg (2013). https://doi.org/10.1007/978-3-642-56026-2
25. Issakhov A., Mashenkova, A.: Numerical study for the assessment of pollutant dispersion from a thermal power plant under the different temperature regimes. Int. J. Environ. Sci. Technol. 1–24 (2019). https://doi.org/10.1007/s13762-019-02211-y
26. Issakhov, A., Bulgakov, R., Zhandaulet, Y.: Numerical simulation of the dynamics of particle motion with different sizes. Eng. Appl. Comput. Fluid Mech. **13**(1), 1–25 (2019)
27. Chung, T.J.: Computational Fluid Dynamics. Cambridge University Press, Cambridge (2002)
28. ANSYS Fluent Theory Guide 15, ANSYS Ltd. (2013)
29. Ajersch, P., Zhou, J.M., Ketler, S., Salcudean, M., Gartshore, I.S.: Multiple jets in a crossflow: detailed measurements and numerical simulations. In: International Gas Turbine and Aeroengine Congress and Exposition, ASME Paper 95-GT-9, pp. 1–16, Houston, TX (1995)
30. Keimasi, M.R., Taeibi-Rahni, M.: Numerical simulation of jets in a crossflow using different turbulence models. AIAA J. **39**(12), 2268–2277 (2001)

Applying Data Assimilation on the Urban Environment

Z. T. Khassenova[✉] and A. T. Kussainova

L.N. Gumilyov Eurasian National University, Astana, Kazakhstan
zthasenova@mail.ru

Abstract. The safety on the urban environment is the most urgent problem of the modern world. A promising avenue for solving this problem is the development of effective monitoring systems, whose mathematical support is based on the application of numerical algorithms for modeling the spread of a pollutant, that evaluate the state of the system in real time [1, 2]. In data assimilation tasks, it is required to predict the value of the model state function in accordance with available observational data, that is, to estimate the "real" state of the system using a mathematical model, a priori information and measurement data. The variational principle of constructing numerical schemes is used in this paper [1, 3]. Numerical experiments were carried out.

Keywords: Data assimilation algorithms · Variational principle · Monitoring · Numerical scheme · Testing

1 Introduction

Providing comfortable living conditions of society and people health in the urban environment is an urgent problem of the modern world. A promising solution to this problem is the development of effective monitoring systems, the mathematical support of which is based on the use of numerical algorithms for modeling the distribution of pollutants that take into account the chemical composition of the atmosphere for preparing decisions on management of quality control of the atmosphere. For these purposes, it is necessary to solve the tasks of assimilation data, where it is required to forecast the value of the model state function in accordance with the available observational data, that is, to estimate the "real" state of the system using of mathematical model, priori information and measurement data. In this direction, special attention is paid to algorithms that can evaluate the state of the system in real time. In the work the variational principle of constructing numerical scheme [1,3] is applied and presented testing of algorithms.

A review of scientific research on this subject has shown that there are three main approaches to the tasks of data assimilation: - variational; - dynamic stochastic (based on the Kalman filter algorithm, particle filters); - hybrid Envar.

Y. Shokin and Z. Shaimardanov (Eds.): CITech 2018, CCIS 998, pp. 125–134, 2019.
https://doi.org/10.1007/978-3-030-12203-4_12

Research of data assimilation tasks began in the 60s of the last century. One of the first works on variational harmonization of data is noted by work Sasaki [4], in which the author proposes a variational approach to solving the tasks of analysis. The basis for the construction of algorithms is the variational principle. Then Kalman's optimization technique was developed. The algorithm of assimilation based on the Kalman filter naturally generalizes the assimilation systems, which are a cycle of forecast-analysis [5, 6]. In works [7–19] a large review is given on the applicability of the Kalman algorithm and the ensemble Kalman filter in the tasks of data assimilation meteorological and oceanic observations. At present, the application of the Kalman filter theory to the tasks of data assimilation has been investigated by many authors.

In their research, scientists Daescu D., Carmichael G., Trevisan A., Uboldi F. discuss new trends in the expand of data assimilation techniques and directed monitoring for improving air quality [20–22]. In the present form variational data assimilation algorithm with direct and adjoint problems first appeared in the paper [23], which was cited in the work [24] by French mathematicians ten years later. More detailed overview can be found in [25]. It should be noted the French mathematical school, which made a great contribution to the theory and practice of the variational data assimilation [24]. Works by Elbern H., Strunk A., Schmidt H., Talagrand O., Courtier P., Th'epaut J.N. and Hollingsworth A. devoted to the implementation of 4D-Var variational methods. Also academician G. I. Marchuk developed a four-dimensional variational method, which is based on perturbation theory of conjugate equations for discrete models of atmospheric and ocean dynamics [26]. Various applications of variational principles for constructing numerical schemes for nonlinear mathematical models, methods of the theory of model sensitivity, and data assimilation of measurement from various observational systems are described in the monograph [1].

The main results of scientific work in this field by the Novosibirsk scientists V.V. Penenko, A.V. Penenko, A. Baklanov, E.A. Tsvetovoy, E.A. it is the development of the theory and methods for solving interrelated tasks of ecology and climate, the transfer and transformation of pollutants in the atmosphere, environmental prediction and design based on variational principles [27–33]. With the help of the proposed models, scientists solved the applied tasks of problems environmental monitoring for cities such as Moscow, Novosibirsk, Tomsk, Sofia, Ulan-Ude, Astrakhan, and Alma-Ata; Ust-Kamenogorsk.

A lot of research was carried out in his works by scientist Lorenc A.C. about the management of 3D and 4D-Var variational principles in data assimilation, the Kalman ensemble and its potentials, methods of analysis of numerical weather forecasting, modeling and development schemes of operational variational data assimilation, which are currently actively developing.

Research and creation of the theoretical and methodological basis of information technology of assessment and prediction of the state of the environment, taking into account the impact on the level of pollution of hazardous natural and man-made processes are devoted to the work of such domestic authors as, Danaev N.T., Temirbekov N.M., Zakarin E.A., Sultangazin U., Bakirbaev B., Aidosov A. and others.

Environmental monitoring in the example of the city of Ust-Kamenogorsk of the East Kazakhstan region is also carried out by domestic scientists: Temirbekov, Madiyarov, Malgazhdarov, Abdoldina, Turganbaev, Soltan, Rakhmetullina, Denisova [34–38].

Thus, in practice, two main types of assimilation systems are used: continuous and discrete. If in continuous systems of observations is carried out at a time, then in discrete - at times corresponding to the main synoptic terms of observations. On the basis of the complexity of the implementation, F. Bouttier and P. Courtier presented a conditional classification of data assimilation algorithms [39].

In the last decade, the search for effective algorithms for spatio-temporal data assimilation in the context of with reference to atmospheric chemistry tasks is increasing. An overview of the main approaches to solving this problem is given in work [40].

The general scheme of data assimilation can be represented as follows, as shown in Fig. 1.

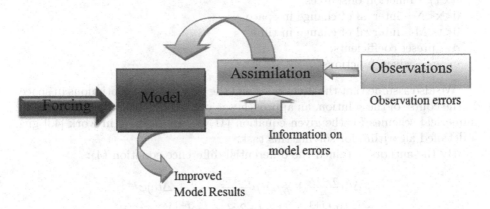

Fig. 1. The general scheme of data assimilation

Extensive research by the world and domestic scientists on the algorithms and methods of distribution of pollutants, including industrial cities, shows an actual global problem that requires the development of further research in this area and the solution of tasks for the benefit of a comfortable and healthy life of society with the introduction of modern high-performance computing systems.

2   Mathematical Support

Model of the Process of Impurity Transfer in the Atmosphere

As a model of the process of impurity transfer in the atmosphere, the boundary value problem for the nonstationary transport equation with diffusion on the bounded interval $(0, X)(0, 1)$ over the space is considered:

$$\frac{\partial}{\partial x}\mu\frac{\partial \varphi}{\partial x} - u\frac{\partial \varphi}{\partial x} = \frac{\partial \varphi}{\partial t} + c\varphi - f(x,t) \tag{1}$$

with the following boundary conditions:

$$\begin{aligned}-\mu\frac{\partial \varphi}{\partial x} + a\varphi = q_L, & \quad x = 0 \\ \mu\frac{\partial \varphi}{\partial x} + a\varphi = q_R, & \quad x = X\end{aligned} \tag{2}$$

and initial data:

$$\varphi = \varphi_0, t_0 = 0 \tag{3}$$

where

φ – function of impurity concentration,
φ_0 – initial distribution of concentration,
μ_0 – coefficient of turbulent exchange,
u – impurity transfer rate,
c – coefficient disintegration,
f(x,t) – function of sources,
$0 < x < N$ – interval of change in space,
$0 < t < M$ – interval of change in time,
α – preset coefficients,
q_L, q_R – given functions.

We also assume that the function φ and the unit $\mu\frac{\partial \varphi}{\partial x}$ are continuous in space. In the course of the solution, an approach was used to construct discrete-analytic numerical schemes for the given equation [41, 42]. By authors in work [43] give a detailed algorithm for solving this task.

By the authors obtained the differential-difference equation (4):

$$\begin{aligned}-\Delta tu\frac{\partial \varphi^{j+1}}{\partial x} + \frac{\partial}{\partial x}\Delta t\mu\frac{\partial \varphi^{j+1}}{\partial x} - (\tfrac{3}{2} + \Delta tc)\varphi^{j+1} \\ = -\Delta tf^{j+1}(x,t) + (-2\varphi^j + \tfrac{1}{2}\varphi^{j-1}),\end{aligned} \tag{4}$$

To construct discrete approximations, the authors used the integral identity obtained after multiplying (4) by a sufficiently smooth conjugate function φ^* and two integrations by parts (5):

$$\int_{x_{i-1}}^{x_i}(-\frac{\partial(\Delta tu)\varphi^*}{\partial x} + \frac{\partial}{\partial x}\Delta t\mu\frac{\partial \varphi^*}{\partial x})\varphi^{j+1}dx + \Delta tu\varphi\varphi^*|_{x_{i-1}}^{x_i} + \Delta t\mu\frac{\partial \varphi}{\partial x}\varphi^*\Big|_{x_{i-1}}^{x_i} - \Delta t\mu\frac{\partial \varphi^*}{\partial x}\varphi\Big|_{x_{i-1}}^{x_i}$$
$$- \int_{x_{i-1}}^{x_i}((\tfrac{3}{2} + \Delta tc)\varphi^{j+1} + \Delta tf^{j+1}(x,t) - (-2\varphi^j + \tfrac{1}{2}\varphi^{j-1}))\varphi^*dx = 0 \tag{5}$$

Variational Data Assimilation Algorithms of Observations of Point Measurements of Impurity Concentration in Real Time

In data assimilation tasks, it is forecasted the value of the model state function in accordance with available observational data. Using the algorithm proposed in work [43], we studied data assimilation algorithms of observation data for point measurements of impurity concentration in real time. In works [44, 45] the latest research and results in this direction are presented.

The inverse modeling tool is the uncertainty function. The concentration values were measured in some finite set of points in space and in time. To continue the concentration field from the observation points, we use model (1) with boundary conditions (2) and an algorithm for solving the tasks sequential variational data assimilation algorithms [44]. Denote by I_i^{j+1} the result of measurements at the j-th time moment at the grid point with index i, M_i^{j+1} - mask of the measurement system.

For approximate the a priori model, we use an implicit scheme by adding right side function of control ξ. As a result, $j{+}1$ s time step we get:

$$
\begin{aligned}
-A_i\varphi_{i+1}^{j+1} + B_i\varphi_i^{j+1} &= \varphi_i^j + \Delta t\xi_i^{j+1}, \ i = 0,\\
-A_i\varphi_{i+1}^{j+1} + B_i\varphi_i^{j+1} - C_i\varphi_{i-1}^{j+1} &= \varphi_i^j + \Delta t\xi_i^{j+1}, \ i = 1,\ldots,N-2,\\
B_i\varphi_i^{j+1} - C_i\varphi_{i-1}^{j+1} &= \varphi_i^j + \Delta t\xi_i^{j+1}, \ i = N-1.
\end{aligned}
$$

The model was adjusted in accordance with the incoming observations I_i^{j+1}. To do this we minimize the functional:

$$\Phi(\varphi^{j+1},\xi^{j+1}) = \sum_{i=0}^{N-1}\left(\varphi_i^{j+1} - I_i^{j+1}\right)^2 M_i^{j+1}\Delta t + \sum_{i=0}^{N-1}\left(\xi_i^{j+1}\right)^2\Delta t.$$

Further, the authors wrote out an extended functional

$$\overline{\Phi}(\varphi^{j+1},\xi^{j+1},\psi) = \left(\sum_{i=0}^{N-1}\left(\varphi_i^{j+1} - I_i^{j+1}\right)^2 M_i^{j+1} + \sum_{i=0}^{N-1}\left(\xi_i^{j+1}\right)^2\right)\Delta t$$

$$+ \sum_{i=0}^{N-1}\left(-\Delta t\xi_i^{j+1} - a_i\varphi_{i+1}^{j+1} + b_i\varphi_i^{j+1} - c_i\varphi_{i-1}^{j+1} - \varphi_i^j\right)\psi_i$$

As a result, a matrix system was obtained:
when $i = 0$,

$$\begin{pmatrix} A_i & 0 \\ 0 & C_{i+1} \end{pmatrix}\begin{pmatrix} \varphi_{i+1}^{j+1} \\ \psi_{i+1} \end{pmatrix} + \begin{pmatrix} B_i & -\frac{\Delta t}{2} \\ 2M_i^{j+1}\Delta t & B_i \end{pmatrix}\begin{pmatrix} \varphi_i^{j+1} \\ \psi_i \end{pmatrix} = \begin{pmatrix} \varphi_i^j \\ 2M_i^{j+1}\Delta t I_i^{j+1} \end{pmatrix} \quad (6)$$

when $i = 1,\ldots,N-2$,

$$-\begin{pmatrix} A_i & 0 \\ 0 & C_{i+1} \end{pmatrix}\begin{pmatrix} \varphi_{i+1}^{j+1} \\ \psi_{i+1} \end{pmatrix} + \begin{pmatrix} B_i & -\frac{\Delta t}{2} \\ 2M_i^{j+1}\Delta t & B_i \end{pmatrix}\begin{pmatrix} \varphi_i^{j+1} \\ \psi_i \end{pmatrix} - \begin{pmatrix} C_i & 0 \\ 0 & A_{i-1} \end{pmatrix}$$

$$\begin{pmatrix} \varphi_{i-1}^{j+1} \\ \psi_{i-1} \end{pmatrix} = \begin{pmatrix} \varphi_i^j \\ 2M_i^{j+1}\Delta t I_i^{j+1} \end{pmatrix} \quad (7)$$

when $i = N-1$,

$$\begin{pmatrix} B_i & -\frac{\Delta t}{2} \\ 2M_i^{j+1}\Delta t & B_i \end{pmatrix}\begin{pmatrix} \varphi_i^{j+1} \\ \psi_i \end{pmatrix} - \begin{pmatrix} C_i & 0 \\ 0 & A_{i-1} \end{pmatrix}\begin{pmatrix} \varphi_{i-1}^{j+1} \\ \psi_{i-1} \end{pmatrix} = \begin{pmatrix} \varphi_i^j \\ 2M_i^{j+1}\Delta t I_i^{j+1} \end{pmatrix} \quad (8)$$

To solve it, a matrix sweep was applied.

For illustrate the operation of the algorithms, a series of numerical experiments.

Experiment 1. Test experiment. The following options are selected: coefficient of turbulent exchange: $\mu = 0,06\,\mathrm{m^2/s}$, transfer rate: $u = 0.7\,\mathrm{m/s}$. The result of the work is shown in Fig. 2 in the form of graphs.

For the experiment, a mobile surveillance post was chosen, the trajectory of which is shown in Fig. 2b. In Fig. 2(c) the solution of the assimilation problem is adjusted to the measurement data, and at the remaining points the solution "tends" to the exact solution depicted in Fig. 2a.

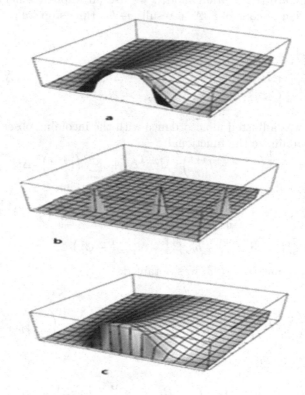

Fig. 2. The result of experiment 1

Experiment No. 2. Test experiment for industrial city of Almaty. According to meteorological data, wind speed $u = 3\,\mathrm{m/s}$, wind direction - western, coefficient of turbulent exchange: $\mu = 0,06\,\mathrm{m^2/s}$. The result of the numerical experiment is shown in Fig. 3.

The results of testing the data assimilation algorithm: the data assimilation procedure may impose additional requirements on the numerical scheme for the basic model of physical processes; the appearance of additional peaks in the solution of the problem of data assimilation in contrast to the solution of a direct problem when combining schemes with a high order of accuracy and time-discontinuous solutions to data assimilation problems.

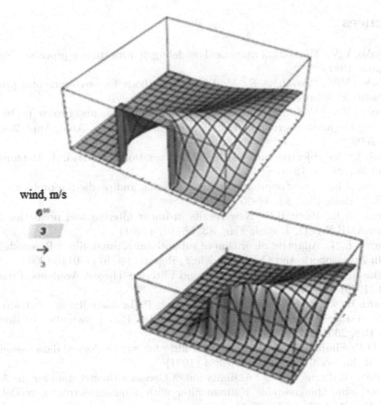

Fig. 3. The result of numerical experiment 2

3   Conclusion

In the course of the study, the existing algorithms of data assimilation used in the works of modern scientists were studied. To improve the solution of the environmental problem for the urban environment, it is required to use these algorithms with more accurate input information, for example, point measurements, complex chemistry with many species, wind direction changes in real time.

Adaptation of data assimilation algorithms of monitoring is as follows:

- the coordinates of the observation points in the algorithm are determined in accordance with the coordinates of the pollution observation posts of city;
- adaptation of the diffusion convection model to the conditions of city;
- chemical composition of the atmosphere on the urban environment.

The work was supported by the Ministry of Education and Science of Republic of Kazakhstan, under the Grant "Development of a new information system and database to optimize monitoring of atmospheric air pollution by heavy metals" (contract number 132, 12 March, 2018).

References

1. Penenko, V.V.: Methods of numerical modeling of atmospheric processes. Hydrometeopub (1981)
2. Penenko, V.V., Aloyan, A.E.: Models and methods for environmental problems. Novosibirsk, Science, Siberian (1985)
3. Penenko, V.V.: Variational methods of data assimilation and inverse problems for studying the atmosphere, ocean, and environment. Numer. Anal. Appl. **2**(4), 341–351 (2009)
4. Sasaki, I.: An objective analysis based on variational method. J. Meteorol. Soc. Japan **36**, 29–30 (1958)
5. Kalman, R.E.: A new approach to linear filtering and prediction problems. Trans. AME. J. Basic Eng. **82**, 34–35 (1960)
6. Kalman, R.E., Bucy, R.S.: New results in linear filtering and prediction theory. Trans. AME Ser. D. J. Basic Eng. **83**, 95–107 (1961)
7. Klimova, E.G.: Adaptive algorithm of suboptimal Kalman filter. Research Activities in Atmospheric and Ocean Modeling, Report, pp. 0117–0118 (2004)
8. Jazwinski, A.H.: Stochastic Processes and Filtering Theory. Academic Press, New York (1970)
9. Menard, R., Cohn, S.E., Chang, L.-P., Lyster, P.M.: Assimilation of stratospheric chemical tracer observations using a Kalman filter. Part 1: formulation. Mon. Wea. Rev. **128**, 2654–2671 (2000)
10. Dee, D.P.: Simplification of the Kalman filter for meteorological data assimilation. Q. J. R. Meteorol. Soc. **117**, 365–384 (1991)
11. Evensen, G., Leeuwen, P.J.: Assimilation of Geosat altimeter data for the Agulhas current using the ensemble Kalman filter with a quasigeostrophic model. Mon. Wea. Rev. **124**, 85–96 (1996)
12. Heemink, A.W., Verlaan, M., Segers, A.J.: Variance reduced ensemble Kalman filtering. Mon. Wea. Rev. **129**, 1718–1728 (2001)
13. Miyoshi, T., Sato, Y.: Assimilating satellite radiances with a local ensemble transform Kalman filter (LETKF) applied to the JMA global model (GSM). SOLA **135**, 37–40 (2007)
14. Miyoshi, T., Yamane, S.: Local ensemble transform Kalman filter with an AGCM at a T159/L48 resolution. Mon. Wea. Rev. **135**, 3841–3861 (2007)
15. Takemasa, M., Yoshiaki, S., Takashi, K.: Ensemble Kalman filter and 4D-Var intercomparison with the Japanese operational global analysis and prediction system. Mon. Wea. Rev. **138**(7), 2846–2866 (2010)
16. Whitaker, J.S., Hamill, T.M., Wei, X., Song, Y., Toth, Z.: Ensemble data assimilation with the NCEP global forecast system. Mon. Wea. Rev. **136**, 463–482 (2008)
17. Kalnay, E., Park, S.K., Pu, Z., Gao, J.: Application of quasi-inverse method to data assimilation. Mon. Wea. Rev. **128**, 864–875 (2008)
18. Rabier, F.: Extended assimilation and forecast experiments with a four-dimensional variational assimilation system. Q. J. R. Meteorol. Soc. **124**, 1861–1887 (1998)
19. Mitchel, H.L., Houtekamer, P.L.: An adaptive ensemble Kalman filter. Mon. Wea. Rev. **128**, 413–433 (2000)
20. Daescu, D., Carmichael, G.: An adjoint sensitivity method for the adaptive location of the observations in air quality modeling. J. Atmos. Sci. **60**, 434–449 (2003)
21. Trevisan, A., Uboldi, F.: Assimilation of standard and targeted observations within the unstable subspace of the observation-analysis-forecast cycle system. J. Atmos. Sci. **65**, 103–113 (2004)

22. Palatella, L., Carrassi, A., Trevisan, A.: Lyapunov vectors and assimilation in the unstable subspace: theory and applications. J. Phys. A: Math. Theor. **46**, 254020 (2013)
23. Penenko, V.V., Obraztsov, N.N.: Variational method of matching fields of meteorological elements. Meteorol. Hydrol. **11**, 3–16 (1976)
24. Le Dimet, F., Talagrand, O.: Variational algorithms for analysis and assimilation of meteorological observations: theoretical aspects. Tellus **38A**, 97–110 (1986)
25. Bocquet, M., et al.: Data assimilation in atmospheric chemistry models: current status and future prospects for coupled chemistry meteorology models. Atmos. Chem. Phys. Discuss. **14**, 32233–32323 (2014)
26. Marchuk, G.I.: Basic and conjugate equations of atmospheric and ocean dynamics. Meteorol. Hydrol. **2**, 17–34 (1974)
27. Penenko, V.V.: Computational aspects of modeling the dynamics of atmospheric processes and evaluating the influence of various factors on the dynamics of the atmosphere. Some Problems of Computational and Applied Mathematics, Novosibirsk, Science, pp. 61–76 (1975)
28. Penenko, V.V.: Variational data assimilation in real time. Comput. Technol. **10**(8), 9–20 (2005)
29. Penenko, V.V.: Forecasting changes in atmospheric quality with estimation of uncertainties from monitoring data. Opt. Atmos. Ocean. **21**, 492–497 (2008)
30. Penenko, V., Tsvetova, E.: Orthogonal decomposition methods for inclusion of climatic data into environmental studies. Ecol. Modell. **217**, 279–291 (2008)
31. Penenko, A.V., Penenko, V.V., Tsvetova, E.A.: Consistent algorithms for data assimilation in atmospheric quality monitoring models based on the variational principle with weak constraints. Sib. J. Comput. Math. **19**, 401–418 (2016)
32. Penenko, V.V., Tsvetova, E.A.: Mathematical models of environmental prediction. Appl. Mech. Tech. Phys. **48**, 428–436 (2007)
33. Pianova, E.A., Faleychik, L.M.: Information-computing equipment for scenarios for assessing the dynamics and quality of the atmosphere. Comput. Technol. **17**, 109–119 (2012)
34. Rakhmetullina, S.Zh., Denisova, N.F., Bitimbayev, I.T.: Application of variational algorithms in the system of ecological monitoring. In: III International Scientific-Practical Conference on Informatization of Society: Mother (2012)
35. Temirbekov, N.M., Madiyarov, M.N., Abdoldina, F.N., Malgazharov, E.A., Temirbekov, A.N.: Mathematical models and information technologies of the atmospheric boundary layer (2011)
36. Madiyarov, M.N.: Geoinformation system for modeling the process of air pollution in the industrial city of the city. Sci. Tech. J. Bull. Eng. Acad. Repub. Kazakhstan **3**(25), 18–23 (2007)
37. Soltan, G.Zh.: Informational and analytical system for decision support in the ecological monitoring of a water body. Vestnik of D. Serikbaev EKSTU **1**, 118–122 (2008)
38. Belginova, S.A., Rakhmetullina, S.Zh., Turganbaev, E.M.: Development of technology for assimilation of environmental monitoring data based on the variational algorithm. In: Materials of the 9th International Asian School-Seminar "Problems of Optimization of Complex Systems", 15–25 August (2013)
39. Bouttier, F., Courtier, P.: Data assimilation concepts and methods. Meteorological Training Course Lecture Series, ECMWF (2002)
40. Elbern, H., Strunk, A., Schmidt, H., Talagrand, O.: Emission rate and chemical state estimation by 4-dimensional variational inversion. Atmos. Chem. Phys. **7**, 3749–3769 (2007)

41. Penenko, V.V.: Numerical schemes for advective-diffusion equations using local problems. Rotaprint of Computing Center SB RAS, Novosibirsk (1990)
42. Penenko, V., Tsvetova, E., Penenko, A.: Variational approach and Euler's integrating factors for environmental studies. Comput. Math. Appl. **67**, 2240–2256 (2014)
43. Penenko, A.V., Kussainova, A.T.: Development of an algorithm for data assimilation for of the model convection-diffusion of an impurity in the atmosphere based on a nonstationary two-layer discrete-analytic numerical scheme. Vestnik of D. Serikbaev EKSTU **2**, 84–91 (2013). ISSN 1561-4212
44. Penenko, A.V.: Some theoretical and applied questions of sequential variational assimilation of data. Comput. Tech. **11**, 35–40 (2006)
45. Penenko, A., Mukatova, Z.S., Penenko, V.V., Gochakov, A., Antokhin, P.N.: Numerical study of the direct variational algorithm of data assimilation in urban conditions. Atmos. Ocean Opt. **31**, 456–462 (2018)

Computer Technologies for Gestures Communication Systems Construction

Iu. Krak$^{(\boxtimes)}$

Taras Shevchenko National University of Kyiv, Kyiv, Ukraine
krak@univ.kiev.ua, Yuri.krak@gmail.com

Abstract. The approach to the development of informational computer technologies for the interactive learning of sign language which use 3D model human expart of sign language is proposed. Methods for modeling fingerspelling information and it's recognizing uses convolution neural networks are created. The implementation of these methods under the operating system Microsoft Windows and on the cross-platform technology have been developed. Methods and software for gestures modeling base on 3D human model and using Motion Capture technology are created.

Keywords: Sing language · Modeling · Recognition · Deaf people · 3D design

1 Introduction

Currently, the society tries to solve the problem of integration of disabled people. The same is true in the relations to the people with hearing impairment and the deaf-mutes. As a rule, sign language (SL) remains the main way of communication among those people [1]. In terms of SL, the movements of body and hands, face, eyes code the information, which is perceived visually.

Therefore the fundamental peculiarities of the SL are that the leading role in the communication belongs to the settings and the elements of gestures are performed and perceived simultaneously.

Gesture is the main unit of the SL. There are three parameters necessary for the description of gesture structure [2]: relations between the place of gesture performance and the body of a communicator, design of a hand demonstrating the gesture; trajectory of the hand. Today, existing systems for learning the sign language are oriented, mainly, to video recording of gestures. The efficient solution of the problem of SL investigation requires the creation of the technologies responsible for the computerization of this process. In fact, it is quite possible to perform such tasks due to the accessibility and modernity of the approach, considering pace of the development of graphic processors and technologies for the 3D graphics. As noted in the paper [3], development of the human-like avatars for gestures modeling with applicatications to creation of the SL teaching systems are very important and perspective. From the other hand avatar-oriented

© Springer Nature Switzerland AG 2019
Y. Shokin and Z. Shaimardanov (Eds.): CITech 2018, CCIS 998, pp. 135–144, 2019.
https://doi.org/10.1007/978-3-030-12203-4_13

approach to SL modeling has general character and can be use to realization any sign language. In this article it is proposed to develop the approaches given in the papers [4–6], and expand them for modeling based on 3D model of fingers of the human hand and the recognition of signs of the alphabet. Develop a system of sign communication based on 3D human model and implement it in the form of the cross-platform information technology.

2   Modeling and Recognition of the Sign Language Fingerspelling Alphabet

Dactyl (fingerspelling) SL is an integral part of sign communication systems and to show individual letters (dactylemes) of the alphabet is designed. Note, that for the main languages of the world the dactylemes are demonstration with the fingers of one human hand (right hand as rule). Necessary create mathematical methods and computer technologies for modeling and demonstration as separative dactylemes as well as continuous sequence of dactylemes for words showing. For given problems solving, investigations were conducted and base on it a new methods, algorithms and models of computer synthesis of fingerspelling alphabet were and effective broadcast of gestures animation via the Internet were created.

To create a software tools, a mathematical model of a simplified human skeleton was used. The hand is represented as a hierarchical structure of bones forming an acyclic directed graph. As mathematical model for fingerspelling alphabet modeling was used of a simplified human hand skeleton, were the hand is represented as a hierarchical structure of bones forming an acyclic directed graph. Based on the proposed mathematical model, an information model that contains a set of fingers vertices in the initial state, a sets of indices for representing triangles, a set of normals for each vertex, a hand texture coordinates describing the surface of the hand was created [4]. Hand visualization occurs base on through the spinning procedure, which it the most effective method of animation [4].

The developed methods and algorithms were implemented in the form of information technology (see Fig. 1), were indicated numbers mean next activities: 1 – area of displaying dactylemes alphabet; 2 – panel of displaying playback progress of dactylemes or words; 3 – input panel for words; 4 – list of letters of the fingerspelling alphabets; 5 – button «spell», the process of fingerspelling of input word begins when the button is clicked; 6 – panel to demonstrate the verbal description of a hand configuration that correspond to the current displayed dactylemes; 7 – panel to display written letter and a picture that correspond to the current displayed dactylemes; 8 – indicator of a location of a hand rotation; 9 – define the pace of fingerspelling [5].

It should be noted that on the basis of the created information technology software tools can be developed for the simulation of any dactyal sign language from a one-handed finger alphabet. For example, on a Fig. 2 shows the work of the created software for fingerspelling alphabet modeling of the Kazakh sign language.

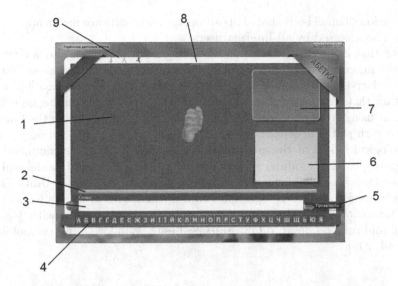

Fig. 1. The main activities of a software for fingerspelling alphabet modelling

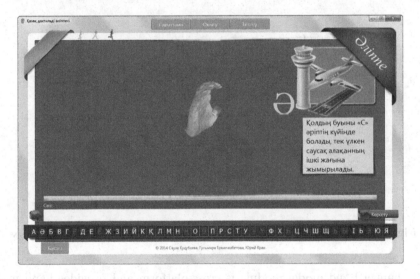

Fig. 2. Fingerspelling alphabet modeling of Kazakh sign language

The created software package runs in the OS Windows environment and over the Internet. It is important to note that the parallelization of frame preparation algorithms allows you to play animation with an increased frame rate when using multi-core processors, thereby increasing the readability of the source files of applications. As advantage of created approach is it transfer a controlled media stream through the Internet with the adaptation to the width of the data transmission channel is created. This allows to create software media applications that can be accessed by all Internet users.

Note that created software was oriented on OS Windows and for running proposed information technology in any operation systems the cross-platform software dactylemes alphabet modeling and recognition was created [6]. It very important that created software should solve the problem of running on existing platforms using cross-platform development without implementing the functionality for each platform separately.

Infological model of the cross-platform technology which is composition of three cross-platform modules is demonstrate in Fig. 3: 3D hand model and user interface (which are implemented with cross-platform framework Unity3D [7]), and software for gesture recognition (implemented with cross platform framework Tensorflow [8]). The main functionality is implemented with C# and Python and runs on desktop OS (MacOS, Linux, Windows) and on mobile OS (Android, iOS).

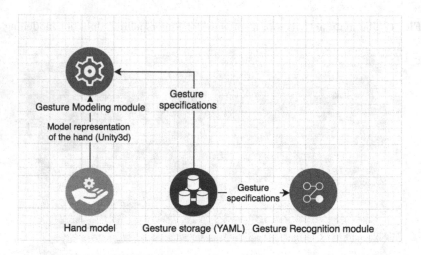

Fig. 3. Infologic model of cross-platform gesture communication technology

3D human hand model module is cross-platform and provides hand model representation for gesture recognition module. Function renderer for human hand receives hand model representation and gesture specifications from gesture storage module, and provides a high-polygon rendered hand model. Module for gesture learning and module for gesture modification are implemented with cross-platform Unity3D, both taking as input results of hand model renderer. Module for gesture modification provides updated gesture specifications and transmits them to gesture storage. Note that for the storage of a gesture, the BVH file format was used. For gesture recognition is proposed to be implemented with Tensorflow framework and receives as input data: model of the 3D human hand, gesture specifications and input from usual camera [6].

Note that for Unity3D framework is able to effectively reproduce a realistic human hand 3D model which consists of more than 70,000 polygons (see Fig. 4). Based on the anatomy of the hand within Unity3D hand model was developed with 25 degrees of freedom, four of them located in the metacarpalcarpal joint, to the little finger and thumb to provide movement palm. The thumb has 5 degrees of mobility, middle and index fingers have 4 degrees of mobility (metatarsophalangeal joint with two degrees of mobility, and the distal and proximal interphalangeal joints each have one) [9].

Fig. 4. 3D model for gestures demonstration under iOS platform

Modules for gesture modeling and recognition are developed with cross-platform tools (frameworks based on Python, C++) can be embedded into information and gesture communication cross-platform technology. Multiple approaches were considered as an approach for gesture recognition [10, 11].

Thus SL recognition tools are consists of taking an input of video stream, extracting motion features that reflect SL linguistic terms, and then using pattern mining techniques or machine learning approaches on the training data. For example, in the paper [9] propose a novel method called Sequential Pattern Mining (SPM) that utilizes tree structures to classify characteristic features.

Convolution neural networks have such advantages [6]:

(1) no need in hand crafted features of gestures on images;
(2) predictive model is able to generalize on users and surrounding not occurring during training;
(3) robustness to different scales, lighting conditions and environment.

As a result of experimental studies F1-score of gesture recognition on test dataset of 0.2 fraction of whole dataset for 100 image samples is equal 0,6, for 200 image samples is equal 0,74, for 500 image samples is equal 0,8 and for 1000 image samples is equal 0,82.

Usage of cross-platform neural network framework such as Tensorflow allows to implement gesture recognition as a cross-platform module of proposed technology and serve trained recognition model on server or transfer it to the device [6].

Created software [6] offered and used in the implementation of information technology is cross-platform and operates unchanged regardless of operating system (Windows, Linux, Android, iOS), CPU type (x86, arm), and the type of hardware (mobile or stationary device).

With its cross-platform build system Unity3D it is possible to create applications for each platform without porting or changing the original code.

As there are no specific hardware requirements for information technology for modeling SL, there are objective obstacles for performance speed of older generations devices. To overcome this problem, the following adaptive approach to information technology was proposed as shown on Fig. 5.

Further modules implementation will leverage from existing cross-platform technology. Modules for gesture learning and recognition, developed with cross-platform technologies (Python, Tensorflow) will be embedded into information and gesture communication cross-platform technology. In case of the mobile app (iOS, Android) or application on the device with a stationary operating system (Windows, Linux), during installation on the device information technology analyses the existing hardware and, depending on its capacity, conducts a series of adjustments.

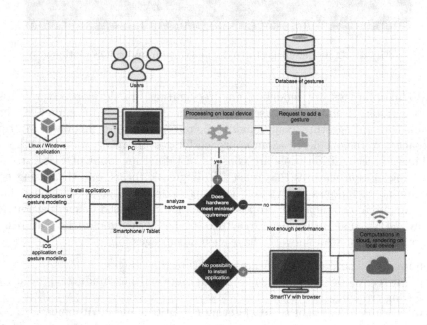

Fig. 5. Cross-platform and adaptive execution of information technology scheme

The effectiveness of the proposed approach is shown in building cross-platform technology for modeling and recognition of Ukrainian fingerspelling (dactyl) alphabet [4,6,11]. Based on this technology, training programs for any one-handed fingerspelling alphabet can be created.

3 Information Technology for Simulation of Sign Language

The general scheme for SL uses for communication with deaf people is shown on a Fig. 6 and will provide the following features [5]:

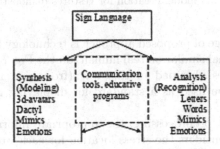

Fig. 6. Sign language communication technology

- a module for translation of the usual text into the SL (text-to-gesture); the module will provide demonstration animation of a common and official SL by presenting the output on a 3D human model;
- mimics and animation (with regard to emotional components) during the pronunciation process;
- lips reading module for recognition of the text being pronounced.

For the implementation of the suggested concept of computer-aided non-verbal communication, a series of research works have been made and the appropriate software has been developed (see, for example papers [2,5,10]). For the 3D model for SL animation synthesis, the geometrical classes of vector-based gestures are described. These classes were formed using Motion Capture (MoCap) technology [12]. MoCap is a technology for retrieving real-world 3D coordinates using multiple video streams recorded from different viewpoints. Then the coordinates are used to determine values in the 3D mathematical model. The key frames are determined by using tracking technology [5,13].

BVH file format for the storage of a gesture was used [5].

Method of the 3D model creation of real human using Motion Capture technology is show on the Fig. 7. Note that the sign language interpreter was used for gestures demonstration (on the Fig. 7 it a left hand picture) and his movements were transferred to the 3D model (on the Fig. 7 it a right hand picture).

Fig. 7. 3D model creation for gestures demonstration

The main advantage of proposed approach is technology for 3D human model creation which maximal similar on real human (on the Fig. 7 it a central picture).

Note that gestures obtained in this way (from sign language interpreter demonstration) can be used as a standard for displaying gestures of a particular sign language.

For the input text preprocessing, the appropriate informational technology was created, which considers the stress location for each word, specifies its normalized word form; contains synonyms and idioms. The model is represented as a set of tables in a relational database along with a set of stored procedures which implement all the required functionality.

For the implementation of visualization and pronunciation feature of a custom text, the appropriate synthesizer has been created. It allows creating the voice equivalent of a custom text using different voices and voice characteristics (volume, distance).

For the complex verification of the suggested technology the appropriate software has been created (Fig. 8). It is used for translation of a custom text into the SL.

The created approach which base on uses the following algorithm for SL synthesis:

(1) a speech equivalent is synthesized for the input text;
(2) the input text is parsed as separative words;
(3) create speech synthesis for the input text;
(4) for each word its normalized form (infinitive) is found by performing a look up in the database;
(5) for each normalized word form a gesture is looked up (represented as a 3D model human movements);
(6) in case the gesture is not found, the word will be shown using fingerspelling alphabet.

The created information technologies was implemented in the specialized schools for deaf people for sign language learning was implemented and effectiveness ones was demonstrated.

Fig. 8. Computer system for SL modelling and learning

4   Conclusion

In the paper suggests a complex approach to creation of the informational technologies for communication with deaf people using sign language: gestures creation, modeling, animation and recognition. Convolution neural networks are used for recognition dactylemes and experiments good recognition results are demonstrated.

As novel approach to the 3D human model creation for sign language demonstration was create model maximal similar on real human. The technology uses 3D models of human body, hands and fingers. Cross-platform technology proposed for solves the problem of execution on the existing multiple platforms without implementing functionality under each platform separately is created. This is one of the important advantages of the proposed approach to simulation and recognition of sign languages.

Thus, in this paper it was shown the effectiveness of the technologies 3D modeling sign language for creation using cross-platform common tools for modeling and recognition sign language. Effectiveness created technologies was demonstrated for modeling and recognition of Ukrainian sign language. Information and gesture communication technology was developed with further scaling capabilities in mind for gestures of other languages such as Polish sing language, Kazakh sign language, English sign language etc.

References

1. Brentari, D. (ed.): Sing Languages. Cambridge University Press, Cambridge (2010)
2. Stokoe Jr., W.C.: Sing language structure: an outline of the visual communication systems of the American deaf. University of Buffalo (1960)
3. Smith, R., Morrissey, S., Somers, H.: HCI for the deaf community: developing human-like avatars for sign language synthesis. In: Proceedings of the 4th Irish Human Computer Interaction Conference, Dublin, pp. 129–136 (2010)
4. Kryvonos, Iu.G., Krak, Yu.V., Barchukova, Yu.V., Trocenko, B.A.: Human hand motion parametrization for dactylemes modeling. J. Autom. Inf. Sci. **43**(12), 1–11 (2011)
5. Krak, Iu., Kryvonos, Iu., Wojcik, W.: Interactive sytems for sign language learning. In: 6th International Conference on Application of Information and Communication Technologies (AICT), pp. 114–116 (2012)
6. Kondratiuk, S., Krak, Iu.: Dactyl alphabet modeling and recognition using cross platform software. In: IEEE Second International Conference on Data Stream Minning & Processing (DSMP), pp. 420–423 (2018)
7. Unity3D framework. [Electronic resource] www.unity3d.com. Accessed 20 Apr 2018
8. Tensorflow framework documentation. [Electronic resource] www.tensorflow.org/api/. Accessed 20 Apr 2018
9. Ong, E.-J., et al.: Sign language recognition using sequential pattern trees. In: IEEE Conference on IEEE Computer Vision and Pattern Recognition (CVPR), pp. 2200–2207 (2012)
10. Kryvonos, I.G., Krak, I.V., Barmak, O.V., Shkilniuk, D.V.: Construction and identification of elements of sign communication. Cybern. Syst. Anal. **49**(2), 163–172 (2013)
11. Krak, Yu.V., Golik, A.A., Kasianiuk, V.S.: Recognition of dactylemes of Ukrainian sign language based on the geometric characteristics of hand contours defects. J. Autom. Inf. Sci. **48**(12), 1–11 (2011)
12. Menache, A.: Understanding Motion Capture for Computer Animation and Video Games. Morgan Kaufmann, Burlington (2000)
13. Avidan, S.: Support vector tracking. In: Proceedings of IEEE Conference on Computer Vision and Pattern Recognition 1, Kauai, Hawaii, pp. 84–191 (2001)

Development of Information Systems for Scientific Research

Yu. Molorodov[✉], A. Fedotov, and D. Khodorchenko

Institute of Computational Technologies of the Siberian Branch of the Russian
Academy of Sciences,
Academician M.A. Lavrentiev avenue, 6, 630090 Novosibirsk, Russia
yumo@ict.sbras.ru, fedotov@sbras.ru, hdaria@yandex.ru
http://www.ict.nsc.ru

Abstract. Based on the conceptual model of the information system, which provides an abstract representation of entities and relationships (connections between entities), support is provided for the architecture of a universal information system related to a particular area of scientific knowledge.

The conceptual model includes the basic entities: actors (persons, actors, organizations and other subjects of activity, including computer applications). An essential component of the conceptual model are documents, publications, dictionary articles, key terms, data and other objects of activity, including facts - a special kind of document. In turn, facts are understood as characteristics of entities described in the ontology of the information system, represented as a single value of data.

Keywords: Information system · Conceptual model · Actors · Essences · Facts

1 Introduction

To date, huge amounts of information have been accumulated and continue to grow in the form of series of data obtained in the course of scientific research. As a rule, they are stored in various paper and electronic publications. A modern approach to processing the data obtained is to digitize and create large digital repositories. This approach allows more flexible organization of access and storage of these data.

To further work with such information repositories, it is necessary to develop specialized Internet resources, the basis of which are conceptual models of information systems for supporting scientific activity. It is urgent, also the problem of clear and convenient display of information, and the interaction of the end user with this information. Therefore, the development of new methods for storing and displaying information does not stop.

The article publcations was supported by RFBR (grant No. 18-07-01457).

Y. Shokin and Z. Shaimardanov (Eds.): CITech 2018, CCIS 998, pp. 145–152, 2019.
https://doi.org/10.1007/978-3-030-12203-4_14

One of the new methods is ontology. They are used as a means of combining information from several sources in one area of knowledge. The creation of ontologies is popular in any subject area. They allow you to structure the data, exchange research results among different researchers. We have worked out convenient methods for entering new data and changing, if necessary, existing ones.

1.1   A Domain Choosing

A standard approach to the systematization of information is the classification of documents using taxonomies. Taxonomy is a subject (thematic) classification that groups terms in the form of controlled dictionaries (thesauri) and organizes them in the form of hierarchical structures.

To describe a particular domain, a certain set of key terms is usually used, each of which designates or describes a concept from a given subject area. The basis of the classification is the separation of concepts (key terms), the establishment of paradigmatic relations between them (for example, the parent-child type) and the comparison of the analyzed document to the highlighted concepts.

The development of specialized thesauri is relevant both for the development and systematization of the conceptual apparatus of the domain (in this case, computer science) and for the logical search for information in full-text databases on the Internet, acting as a means of forming a search requirement, formulating search requirements and adequate automatic indexing, classification of documents.

The formalization of the semantics of the domain in the form of ontology serves not only the goals of a compact and consistent description, it also forms a conceptual basis for the representation of the whole body of knowledge about it. For example, in the system of information support of scientific activity in terms of ontology, the semantics of the data and information resources used in it can be described.

Using the approaches described above, we will demonstrate on bioinformatics, and specifically - its section associated with tick-borne danger in the Altai, Kazakhstan and Siberia. Factographic material obtained during field work is the framework of the future system. At the same time, the basis is the data on mites, the areas of their spread, the pathogens and diseases they are carrying. Additional information will be the actors—persons associated with these mites - pioneers, people studying this species or collecting statistics about it, and others. In this case, the use of ontology as a navigation system allows not only to display information in the form of a clear structure of relationships, but also to supplement the available data with a distributed method with the correspondence to the existing axioms.

2   Domain Ontology

The formalization of the structure of the obtained information about ticks is based not on the taxonomy of ticks. For this, we used the systematized

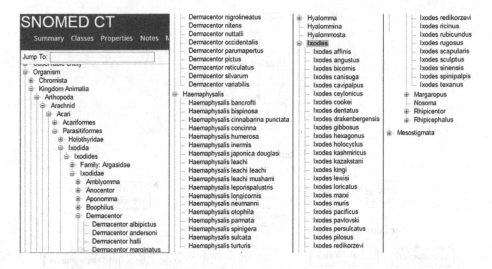

Fig. 1. Of the ixodid ticks taxonomy.

medical nomenclature SNOMED CT [1] and the classification of organisms of the National Center for Biotechnological Information [2,3] (Fig. 1).

Let us single out those species of ixodid ticks that inhabit the regions of Altai, Kazakhstan and Siberia. Information about them will be supplemented by current statistics of bites and infections, as well as by analysis and possible prediction of the situation (Fig. 2).

Different types of ticks can contain different pathogens or groups of pathogens. Many tick-borne diseases show almost identical early symptoms, it can be difficult for health professionals to confidently assess and control the patient who has been bitten by a tick. In addition, in some cases, one tick can be infected, and simultaneously transmit more than one type of pathogen, which further complicates the clinical picture. Knowing the types of ticks and possible infections can alert the physician to specific diseases, thereby facilitating the appropriate diagnosis and treatment. In addition, the ability to better clarify the real threat posed by the tick bite can reduce the unnecessary prescription of antibiotics.

To ensure effective prevention measures, a spatial and temporal analysis of the distribution of ticks, including those infected by one or other pathogens, is necessary. The lack of an effective technology for early detection of known and new pathogens and predicting their spread is one of the important and acute problems. And in this regard, one of the most promising ways to control infectious agents can be continuous surveillance systems. The first step to creating such a system is spatial and temporal analysis based on geoinformation technologies.

Geoinformation system (geographic information system, GIS) is a system for collecting, storing, analyzing and graphically visualizing spatial (geographic) data and associated information about the required objects. The concept of

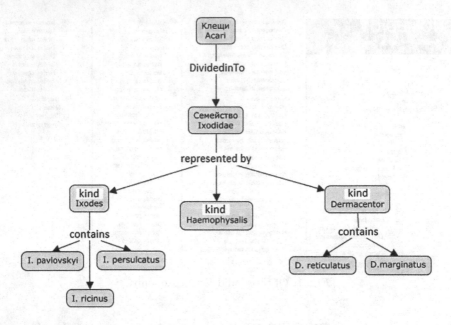

Fig. 2. Of ticks Classification.

geoinformation system is also used in a narrower sense - as a tool (software product) that allows users to search, analyze and edit both a digital map of the terrain and additional information about objects.

In our country, for the effective organization and management of a large volume of accumulated thematic data on the spread of ixodid mites, the cases of the circulation of people with tick bites are used as single databases so the open database "GenBank" [4], in which all registered sequences are stored. This database is publicly available and contains all the annotated DNA and RNA sequences, as well as the sequences of proteins encoded in them. GenBank is supported by the US National Center for Biotechnology Information, part of the National Institutes of Health in the United States, and is available free of charge to researchers around the world. GenBank receives and consolidates data from different laboratories for more than 100,000 different organisms.

3   Of the Information System Architecture

The conceptual basis of the information system of the created Internet resource (knowledge portal) is the ontology described above. Portal ontology introduces formal descriptions of domain concepts in the form of object classes and relations between them, thereby setting up structures for representing real objects and their relationships. In accordance with this, the data on the portal are represented in the form of a semantic network, i.e. as a lot of different types of interconnected information objects. Substantial access to systematic knowledge and information resources is provided through an information system (IS) that provides advanced navigation and search facilities. The architecture of IS is defined by its components, their functions and interaction.

The system is built on the basis of client-server technology and consists of the client part, the server part, and the MySQL database. The information resource is located at (http://ixodes.ict.nsc.ru). The structure of this resource, created on the basis of the ontological campaign is presented in Fig. 3.

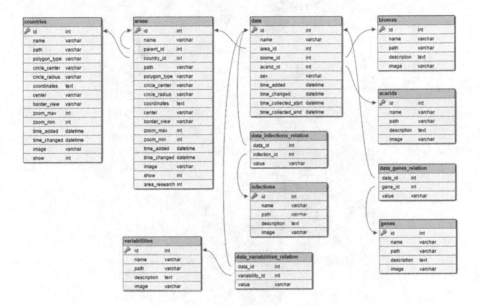

Fig. 3. Database structure.

In the described architecture, the client is the browser (user), and the server is the web server. The logic of the system is distributed between the server and the client, the data is stored in the MYSQL database, information is exchanged through the network. An important advantage of this approach is the fact that customers do not depend on the specific operating system of the user, therefore, the system is a cross-platform service.

At the heart of the client part of the application built on the basis of the structure is the use of cartographic service based on the library Leaflet. This is an innovative, open source, JavaScript library. With it, a vector map is displayed, which is loaded from the MapBox service (https://www.mapbox.com/). The MapBox service provides a wide range of different maps, which can differ in design, language of headings on the map and other parameters. As sources, to generate accurate maps, MapBox uses the OpenStreetMap service (https://www.openstreetmap.org/). OpenStreetMap is a non-commercial, open source project that provides accurate coordinates for all the objects that are on the map. The service uses OpenStreetMap to determine the polygons of areas that form an intuitive interface for the user.

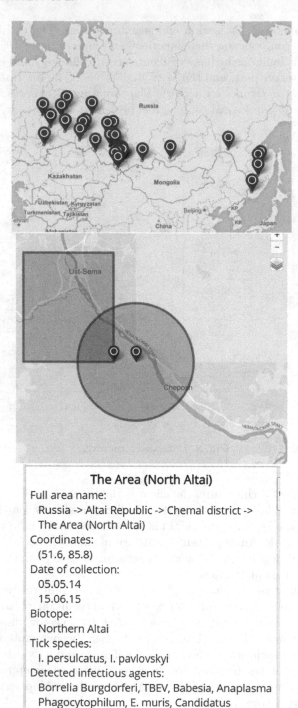

The Area (North Altai)

Full area name:
  Russia -> Altai Republic -> Chemal district ->
  The Area (North Altai)
Coordinates:
  (51.6, 85.8)
Date of collection:
  05.05.14
  15.06.15
Biotope:
  Northern Altai
Tick species:
  I. persulcatus, I. pavlovskyi
Detected infectious agents:
  Borrelia Burgdorferi, TBEV, Babesia, Anaplasma
  Phagocytophilum, E. muris, Candidatus
  Rickettsia Tarasevichiae, SFG - Other Rickettsia
  spp

Fig. 4. On the geographical map type of the information.

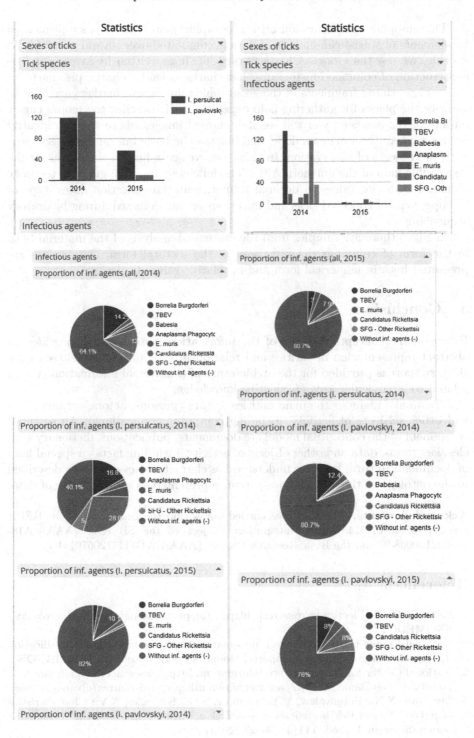

Fig. 5. Of the mathematical processing variants.

The mapping of information on a geographic map (Fig. 4) is supplemented by elements of statistical analysis of data accumulated over several years [5]. To do this, we use the Google Charts library [6]. It is written in JavaScript, the construction of combined histograms, bar charts, calendar charts, pie charts.

In Fig. 4 shows fragments of the geographic map, where markers are used to identify the places for gathering field expeditions. If you click the mouse cursor on any of the markers, you will see its enlarged image, where you can identify the area where the work was done, and below - the basic information associated with the objects of observation. In this case, we see a link to the object on the geographic map of the northern Altai. The following are distinguished: territory names, coordinate values in decimal format, material collection dates, type of biotope, type of collected ticks and infectious agents detected during laboratory sequencing.

In Fig. 5 there are samples from the statistical analysis of the material both in the form of columnar diagrams and in the sectoral form. Data values are presented both in numerical form and in relative values in percent.

4   Conclusion

Based on the conceptual model of the information system, which provides an abstract representation of entities and relationships (connections between entities), support is provided for the architecture of a universal information system related to a particular area of scientific knowledge.

The model includes the main entities: actors (persons, actors, organizations and other subjects of activity, including computer applications). An essential component of the conceptual model are documents, publications, dictionary articles, key terms, data and other objects of activity, including facts - a special kind of document. In turn, facts are understood as characteristics of entities described in the ontology of the information system, represented as a single value of data.

Acknowledgments. The work was carried out within the framework of the RFBR grant (No. 18-07-01457), the Integration Project of the SB RAS (AAAA-A18-118022190008-8) and the Basic Research Project (AAAA-A17-117120670141-7).

References

1. SNOMED CT [Electronic resource]. https://bioportal.bioontology.org/ontologies/SNOMEDCT/
2. National Center for Biotechnology Information (NCBI) Organismal Classification [Electronic resource]. https://bioportal.bioontology.org/ontologies/NCBITAXON
3. National Center for Biotechnology Information. http://www.ncbi.nlm.nih.gov
4. Database "GenBank". https://www.ncbi.nlm.nih.gov/sites/entrez?db=nucleotide
5. Livanova, N.N., Borgoyakov, V.Y., Livanov, S.G., Fomenko, N.V.: Characteristics of natural foci of tick borreliosis Novosibirsk scientific center and the Novosibirsk region. Siberian J. Med. **111**(4), 20–23 (2012)
6. Library Google Charts. https://developers.google.com/chart/

Modelling of Technological Processes of Underground Coal Mining

V. Okolnishnikov[1]([⊠]), A. Ordin[2], and S. Rudometov[1]

[1] Institute of Computational Technologies SB RAS,
Academician M.A. Lavrentiev Avenue, 6, 630090 Novosibirsk, Russia
okoln@mail.ru
[2] N.A. Chinakal Institute of Mining SB RAS,
Krasnyi Avenue 54, 630091 Novosibirsk, Russia
ordin@misd.nsc.ru

Abstract. The paper presents mathematical and simulation models of the technological process of underground coal mining in one of the coal mines in Kuznetsk Coal Basin. The simulation model is developed with the help of our own simulation system MTSS, which contains a specialized library of mining equipment models and a coal seam model. The main goal of the simulation for coal mining technological processes in stoping face is the evaluation of productivity of a cutter-loader depending on different factors. Such factors are: technical parameters of a cutter-loader, size of a longwall face, conditions and constrains of technological processes in stoping face, geophysical state of a coal seam.

Keywords: Coal mining · Simulation system ·
Visual interactive simulation · Longwall mining system

1 Introduction

At present, many coal mines have problems in making decisions to increase productivity, to improve coal production planning, to use new mining machines and new perspective technologies for coal mining. The most suitable way to solve these problems is simulation.

A large number of publications on the use of simulation to support decision making on the design, the development and the optimization of coal mines testifies the importance of these problems [1–8].

To solve these problems the simulation system MTSS [9] was developed. It is a visual interactive and process-oriented discrete simulation system intended to develop and execute the technological processes models. A distinguishing feature of the simulation system is its orientation toward the users who are the experts in a particular subject area (process engineers, mining engineers) not having experience in usage of universal simulation systems. The fast development of models is carried out owing to the visual-interactive interface and specialized libraries of models of technological equipment for specific subject areas.

Y. Shokin and Z. Shaimardanov (Eds.): CITech 2018, CCIS 998, pp. 153–160, 2019.
https://doi.org/10.1007/978-3-030-12203-4_15

The simulation system MTSS provides the user with the following options: visually interactive model construction with a graphical editor, setting model parameters, various modes of model execution, 2D and 3D model visualization. The simulation system MTSS uses 2D as a graphical editor and 2D, 3D for visualization of model execution.

To simulate the technological processes in coal mines in MTSS simulation system the specialized libraries of technological equipment for such coal mine subsystems as an underground conveyor network, a pumping subsystem, and a power supply subsystem were developed. With the use of the specialized libraries a number of models of these subsystems for underground coal mines in Kuznetsk Coal Basin (Russia, Western Siberia) were created [10].

In this article a new specialized library of models of mining equipment in the stoping face is considered. The second section provides a mathematical model of the technological process of coal mining in the stoping face. The third section describes the simulation model of the technological process of coal mining in the stoping face developed using the new specialized library of models and the results of its implementation.

2   The Mathematical Model of Productivity of the Cutter-Loader

The theoretical advance speed of a cutter-loader is [11,12]:

$$V = \frac{30N\eta n_1 K_1}{fP\cos\alpha \pm P\sin\alpha + SDn_2K_2K_3K_4K_5K_6} \tag{1}$$

where

V—the speed of the cutter-loader;
N—the power of the effector motor;
η—the efficiency of the effector reduction gearing;
n_1—the cutting tools in the cutting line;
K_1—the coefficient of the horsepower input to cutter-loader travel;
f—the cutter-loader and transporter friction coefficient;
P—the cutter-loader weight;
α—the angle of inclination of the cutter-loader;
"plus" and *"minus"* in front of the cutter-loader weight specify the cutter-loader movement up and down the longwall, respectively;
S—the weighted average of the coal cutting resistance;
D—the diameter of the augers;
n_2—the cutting tools that synchronously cut the face;
K_2—the coefficient of squeeze which takes into account the cutting force decrease due to the ground pressure;
K_3, K_4, K_5, K_6—the coefficients for the cutting angle, the cutting tool width, the cutting tool dulling and the cutting tool shape, respectively.

Among the above mentioned parameters the coal cutting resistance S influences on the motion rate the most. The coal cutting resistance is considered to be invariable in a certain vast area and is defined by data obtained while drilling the geological prospecting well with (2)

$$S_1 = \frac{k(m_c f_c + m_r f_r)}{m_c + m_r} \tag{2}$$

where

m_c, m_r—the coal mass and the rock mass respectively;
f_c, f_r—the coal hardness and the interbed rock hardness respectively;
k—a certain coefficient.

In the case of several geological prospecting wells S is calculated with the Inverse Distance Weighting method according to (3).

$$S(X,Y) = \begin{cases} \dfrac{\sum\limits_{i=1}^{n} d_i^{-2} S_i}{\sum\limits_{i=1}^{n} d_i^{-2}} & \text{, if } d_i \neq 0 \\ S_i & \text{, if } d_i = 0 \end{cases} \tag{3}$$

where

n—the number of wells nearest to stoping face that are taken into account while calculating;
S_i—the coal cutting resistance in i^{th} well calculated with formula (2);
d_i—the distance between the ith well and the mining face with current position (X,Y), calculated with (4).

$$d_i = \sqrt{(X - x_i)^2 + (Y - y_i)^2} \tag{4}$$

where (x_i, y_i)—the coordinates of the ith well.

While one-way operating of the cutter-loader the time of one-way is:

$$T_{one\,way} = T_1 + T_{scr} + T_{end\,one\,cut} = \frac{L}{V_1} + \frac{L}{V_{scr}} + T_{end\,one\,cut} \tag{5}$$

where

L—the length of the longwall face;
V_1—the speed of the cutter-loader movement up the longwall determined by (1);
V_{scr}—the speed of the cutter-loader movement when scraping operation is performing;
$T_{end\,one\,cut}$—the time of the cutter-loader at the end of the longwall face by one-way operating.

While shuttle operating of the cutter-loader the time of one cycle is:

$$T_{shuttle} = T_1 + T_2 + 2T_{end} = \frac{L}{V_1} + \frac{L}{V_2} + 2T_{end} \tag{6}$$

where

L—the length of the longwall face;
V_1, V_2—the speeds of the cutter-loader movement up and down the longwall determined by (1), respectively;
T_{end}—the time of the cutter-loader at the end of the longwall face by shuttle operating.

While one-way operating of the cutter-loader the productivity of the cutter-loader per time T_{work} is:

$$A_{one\,way} = \frac{\gamma m r L T_{work}}{T_{one\,way}} \tag{7}$$

where

γ—the average mean density of the rock mass;
m—the working-bed height;
r—the cutter-loader cut width.

Substituting V_1 from (1) into (5) and $T_{one\,way}$ from (5) into (7) we get:

$$A_{one\,way} = \frac{\gamma m r T_{work}}{\dfrac{fP\cos\alpha + P\sin\alpha + SDn_2K_2K_3K_4K_5K_6}{30N\eta n_1 K_1} + \dfrac{1}{V_{scr}} + \dfrac{T_{end\,one\,cut}}{L}}. \tag{8}$$

While shuttle operating of the cutter-loader the productivity of the cutter-loader per time T_{work} is:

$$A_{shuttle} = \frac{2\gamma m r L T_{work}}{T_{shuttle}}. \tag{9}$$

Substituting V_1, V_2 from (1) into (6) and $T_{shuttle}$ from (6) into (9) we get:

$$A_{shuttle} = \frac{\gamma m r T_{work}}{\dfrac{fP\cos\alpha + SDn_2K_2K_3K_4K_5K_6}{30N\eta n_1 K_1} + \dfrac{T_{end}}{L}}. \tag{10}$$

3   Integrated Simulating Model of Stoping Operations

In the frames of simulation system MTSS a specialized library of simulating models of mining equipment (a conveyor, a cutter-loader, a self-moving roof support etc.) that are used while coal mining in stoping face was implemented. Using the specialized library of simulating models of mining equipment an integrated model for technological processes of underground coal mining in stoping face was developed. The integrated model involves the following interactive models:

- A coal seam model.
- A cutter-loader model moving up and down the longwall face.
- A model of a self-moving roof support.
- A model of a flight conveyor.

All parameters of the mine equipment models correlate with the parameters of the actual mine equipment operating at one of the coal mine in Kuznetsk Coal Basin. The goals of the simulation for coal mining technological processes in stoping face are:

- The evaluation of productivity of a cutter-loader depending on different factors including the variety of geophysical conditions of the coal seam.
- The input data acquisition (input coal stream) for operating of the belt conveyor network model of the coal mine.

Generally, in the stoping face the following factors influence the productivity of the cutter-loader:

- The geophysical state of the coal seam.
- The advance speed of the cutter-loader.
- The technical characteristics of the cutter-loader.
- The delays at the end of the longwall face.
- The delays associated with the movement of the roof support.
- The delays associated with the flight conveyor.
- The delays associated with the belt conveyor.
- The delays associated with the increase of methane release.
- The delays associated with equipment failure.
- Regulations conditions and maintenance etc.

In this paper the influence of the first six factors on the cutter-loader productivity was studied. Since the main factor influencing on the cutter-loader productivity is the state of the coal seam (the coal cutting resistance) that restricts the advance speed of the cutter-loader, the subject of the research is the detailed simulating of one-way operating and shuttle operating of the cutter-loader together with roof supports movement depending on geophysical state of coal seam. Under these conditions the top speed and the cutter-loader productivity were calculated with (1) and (7) of the mathematical model.

Figure 1 shows the coal seam model with two geological prospecting wells. Figure 2 shows the integrated model of underground coal mining technological processes in stoping face carried out with simulation system MTSS. In Fig. 2 the coal seam with two geological prospecting wells is painted over. The areas with reduced resistance are painted in a lighter tone. The belt conveyor and power lines are designated in the main window.

The equipment parameters and operation modes can be set interactively in the parameters window. The control buttons are entered in the main window: to start the cutter-loader; to stop the cutter-loader.

With the developed simulation model of the technological process of underground coal mining in the stoping face a series of experiments was performed.

Fig. 1. The 3D coal seam model with two geological prospecting wells.

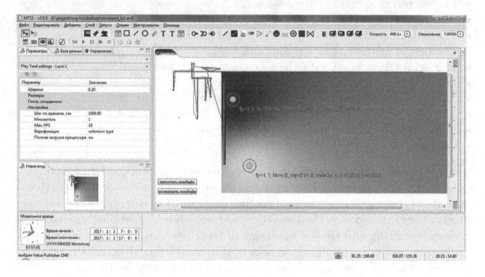

Fig. 2. The main window of the stoping face model.

For one-way and shuttle operations the average productivity of the cutter-loader was calculated, depending on the length of the longwall face. All experiments were carried out under the equal conditions of the passage of the cutter-loader for a certain depth into the coal seam.

The obtained results are presented in the form of graphs in Fig. 3.

The obtained results allow us to conclude the following:

- The shuttle operation of coal mining in the stoping face is more productive in comparison with the one-way operation.
- Increasing the length of the longwall face, beginning from a certain value, does not significantly affect the increase in the productivity of the cutter-loader.

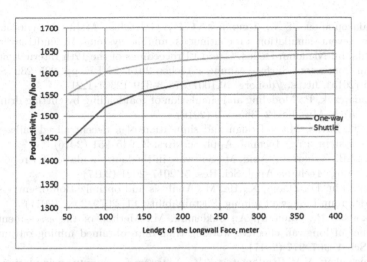

Fig. 3. The dependence of the productivity of the cutter-loader on the length of longwall face.

4   Conclusion

The extension of the considered simulation model of the stoping face operation is assumed to be carried out in the future. Models of aerogasdynamics of methane-air flow in a long stope and models of a belt conveyors subsystem, a ventilating subsystem, and a power supply subsystem of a coal mine will be additionally included into the integrated model of the stoping face operation.

Simulation system MTSS enables engineers with little experience in simulation, investigation and optimization of mining systems to solve complex engineering problems to assess the effectiveness of automation and choice of the optimal technology, which eventually will reduce the costly risks during the design and optimization technological processes in the mining.

The MTSS simulation system can be used not only for simulation of the existing coal mining technologies, but also for simulation of perspective robotized technologies and manless coal mining technologies.

Acknowledgments.. This research was partly financially supported by the Russian Foundation for Basic Research (project 16-07-01179).

References

1. Salama, A., Greberg, J., Schunnesson, H.: The use of discrete event simulation for underground haulage mining equipment selection. Int. J. Min. Miner. Eng. **3**, 256–271 (2014)
2. Fioroni, M., et al.: Logistic evaluation of an underground mine using simulation. In: Proceedings of the Winter Simulation Conference, pp. 1855–1865 (2014)

3. Michalakopoulos, T.N., Roumpos, C.P., Galetakis, M.J., Panagiotou, G.N.: Discrete-event simulation of continuous mining systems in multi-layer lignite deposits. In: Niemann-Delius, C. (ed.) Proceedings of the 12th International Symposium Continuous Surface Mining - Aachen 2014. LNPE, pp. 225–239. Springer, Cham (2015). https://doi.org/10.1007/978-3-319-12301-1_21
4. Gospodarczyk, P.: Modeling and simulation of coal loading by cutting drum in flat seams. Arch. Min. Sci. **2**, 365–379 (2016)
5. Kara, T., Savaş, M.C.: Design and simulation of a decentralized railway traffic control system. Eng. Technol. Appl. Sci. Res. **2**, 945–951 (2016)
6. Ayed, M.B., Zouari, L., Abid, M.: Software in the loop simulation for robot manipulators. Eng. Technol. Appl. Sci. Res. **5**, 2017–2021 (2017)
7. Gao, Y., Liu, D., Zhang, X., He, M.: Analysis and optimization of entry stability in underground longwall mining. Sustainability **11**, 2079–2082 (2017)
8. Snopkowski, R., Napieraj, A., Sukiennik, M.: Method of the assessment of the influence of longwall effective working time onto obtained mining output. Arch. Min. Sci. **4**, 967–977 (2017)
9. Okolnishnikov, V.V., Rudometov, S.V.: A system for computer simulation of technological processes. St. Petersburg State Polytech. Univ. J. Comput. Sci. Telecommun. Control. Syst. **1**, 62–68 (2014)
10. Okolnishnikov, V., Rudometov, S., Zhuravlev, S.: Simulating the various subsystems of a coal mine. Eng. Technol. Appl. Sci. Res. **3**, 993–999 (2016)
11. Ordin, A.A., Metel'kov, A.A.: Optimization of the fully-mechanized stoping face length and efficiency in a coal mine. J. Min. Sci. **2**, 254–264 (2013)
12. Ordin, A.A., Metel'kov, A.A.: Analysis of longwall face output in screw-type cutter-loader-and-scraper conveyor system in underground mining of flat-lying coal beds. J. Min. Sci. **6**, 1173–1179 (2015)

Collection and Processing of Data to Optimize the Monitoring of Atmospheric Air Pollution

Zh. O. Oralbekova, Z. T. Khassenova[✉], and M. G. Zhartybayeva

L.N. Gumilyov Eurasian National University, Astana, Kazakhstan
zthasenova@mail.ru

Abstract. The system of environmental monitoring in Almaty is a collection of data without analysis and processing, without identifying sources of pollution. An urgent solution to this problem is to expand the system of environmental monitoring of the city, since the foothill zone of Almaty is characterized by extremely weak resources for self-cleaning the atmosphere. When developing an information system to optimize the monitoring of atmospheric air pollution by heavy metals, the key stage is the collection, processing and structuring of data [1,2]. The object of research is the air of the surface layer of Almaty, selected at the monitoring posts for the pollution of the air monitoring network. In the course of the research, a selection of sites for air sampling and determination of heavy metals in the samples was carried out. The received data were structured and prepared for inclusion in the created database and further complex mathematical calculations. Technical requirements and limitations to the information system are given.

Keywords: Ecology · Emissions · Heavy metals · Air · System · Monitoring · Database · Database management system · User

1 Introduction

The protection of the air environment is one of the priority directions of the policy of the Republic of Kazakhstan in the field of ecology.

Therefore, gathering and analyzing data on atmospheric air pollution is of particular relevance. The integrated environment will provide the opportunity to process, forecast, produce complex calculations and will help to avoid a large number of emissions of harmful substances into the atmosphere.

The quality of atmospheric air in the city is determined by the emissions of pollutants from the enterprises located on its territory and road transport - the main negative factor that determines the growing pollution of all environments and the concern of citizens [3]. This factor will remain the dominant one in assessing the environmental situation in Almaty.

According to the International Quality of Living Survey, Almaty occupies the 176th place among the 230 largest cities in the world in terms of environmental

Y. Shokin and Z. Shaimardanov (Eds.): CITech 2018, CCIS 998, pp. 161–170, 2019.
https://doi.org/10.1007/978-3-030-12203-4_16

quality [4]. At the same time, among these 230 cities there are also "ten million". According to Almaty Urban Air, some air samples in the "southern capital" showed a degree of air pollution 9 times higher than normal. According to the UN Assembly on the Environment, in the death of more than 25% of children who did not live to 5 years, environmental degradation is responsible.

70% of harmful emissions into the air environment of Almaty fall to the share of transport. According to the Almaty Traffic Police Department, to date, more than 850,000 cars have been registered in the city. When burning fuel, the greatest amount of toxic impurities and heavy metals are released: sulfur oxide, aldehydes, benzopyrene, soot, lead compounds, causing severe diseases, including cancer. Also, studies have shown that the city of Almaty is contaminated mainly with heavy metals, whose high concentration adversely affects the environment.

Given the above problems, we have developed a database in the database management system MS SQL Server for data collection and processing to optimize monitoring of air pollution. In this article we will make the Classification of a priori information, development of attributes of the information under investigation and proceeding from this, we will justify the structure of the created database [5, 6].

2   Rationale for Database Creation

The air quality management system includes state monitoring of ambient air quality, establishment of emission standards, carrying out model calculations of the level of its pollution by atmospheric emissions from industrial enterprises and vehicles. It is being carried out with the aim of developing action plans to reduce the negative impact, as well as carrying out environmental measures aimed at reducing emissions of pollutants into the atmosphere.

The database allows to integrate the available data on the quality of atmospheric air and to conduct their joint analysis [5]. It includes a primary data storage unit.

The sources of primary data are:

- results of automated quality control of atmospheric air;
- data on stationary sources of emissions of pollutants into the atmosphere;
- data on emissions of pollutants into the atmosphere from vehicles.

2.1   Object of Research

The object of the research was air in the surface layer of Almaty, which was selected for 2 monitoring stations for pollution of the air monitoring network: Station No. 1 st. Amangeldy/Ave. Abay and Station No. 2 - Ave. Raimbek/st. Nauryzbai batyr. For our studies, the observation data for these posts monitoring the content of heavy metals were analyzed.

The selection is made at 2 points in the city 1 time per week. The data obtained as a result of laboratory analyzes are used to construct maps of the distribution of concentrations of pollutants - both the main pollutants (heavy metals) and a number of volatile organic compounds [7].

The database uses an internal data format that excludes the ability to import from standard dbf or csv files.

3   Classification of a Priori Information

The information used and generated in the process of air pollution can be divided into the following relatively independent units:

3.1. Information about users and their roles in the system;
3.2. Information on measurements of the concentration of impurities of specific heavy metals in the atmosphere;
3.3. Wind information.

These data can be broken down into four independent databases, or can be combined into one database. The choice depends on the DBMS, on the basis of which the client application will be developed. For example, if you use Oracle, then it is advisable to create a single database. Since at this stage we will use MS SQL Server, which allows simultaneous work with several databases. The following names for databases, tables and fields are working and can be changed. For achievement of the optimization system's objectives of monitoring of pollution of an atmospheric air with heavy metals require a solution of certain problems for optimum performance of a system. In the course of work the main processes of an information system were selected and on the basis of which the corresponding modules are created. A modular principle of creation allows ensuring optimum functioning of an information system. The main components of the system are the following modules that perform basic functions for the system operation:

- "data entry" module;
- meteoparameters module;
- "assimilation of data" module;
- "data analysis" module;
- visualization module.
- "data output" module;

All the main tasks of this system's modules are performed on the server. A web client sends a request to the server which processes a request and replies to the client. (A web client sends server request, the server, in turn, processes a request and again sends to the client.) If there is a necessity, the server sends a request to DB. Each module has the functionality which is presented in Table 1.

3.1   Information About Users and Their Roles in the System - Users

Below we describe the structure of the tables of this database (Tables 2, 3, 4 and 5).

Table 1. Functionality of modules

Module	Functionality of modules
"Data entry" module	Collecting, processing of input data about environmental pollution by heavy metals (concentration of impurity) of the city of Almaty. The input data can be entered and also are imported from various sources
Meteoparameters module	Collecting, processing of meteoparameters (wind speed, a direction of the wind, temperature, atmospheric pressure, etc.)
"Assimilation of data" module	Mathematical support of the environmental monitoring system of atmospheric air with heavy metals is variational algorithms for sequential data assimilation in real time
"Data analysis" module	Definition of the extent of an atmospheric air pollution, a status assessment of a ground layer air of the atmosphere of the city
Visualization module	Visualization of the received results in a graphic view with use of geoinformation technologies
"Data output" module	On the basis of the received results, data output in the form of reports on the extent of pollution of an atmospheric air of the city is made

Table 2. User: FD_AuthUserType

Fields:	Data type	Field description	Field properties
id	bigint	Identifier user	Key field
name_ru	Varchar	Name user	Required (not null)
fullName_ru	Varchar	Full name of the user	Values are allowed NULL
Ord	bigint		Values are allowed NULL
Vis	Integer		Values are allowed NULL

Table 3. User Authentication: AuthUser

Fields	Data type	Field description	Field properties
Id	Bigint	record id	Key field
name_ru	Varchar	Name user	Required (not null)
fullName_ru	Varchar	Full name of the user	Values are allowed NULL
login	Varchar	User login	Required (not null)
passwd	Varchar	Password	Required (not null)
locked	bit	User Status (Lock)	Required (not null)
authUserType	bigint	Authorized User Type	Required (not null)
parent	bigint		Required (not null)
email	Varchar	Mail user	Required (not null)
phoneNumber	Varchar	Phone	Values are allowed NULL

Table 4. Authorized user role: AuthRole

Fields	Data type	Field description	Field properties
Id	Bigint	Identifier id	Key field
name_ru	Varchar	Name user	Required (not null)
fullName_ru	Varchar	Full name of the user	Values are allowed NULL
cmt_ru	text	Comments in Russian	Values are allowed NULL

Table 5. Authorized user role: AuthRoleUser

Fields	Data type	Field description	Field properties
id	bigint	Identifier id	Key field
authRole	bigint	The role of an authorized user	Required (not null)
authUser	bigint	The role of an authorized user	Required (not null)
ord	bigint		Values are allowed NULL
vis	bigint		Values are allowed NULL

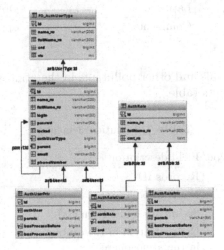

Fig. 1. User unit

The FD_BusProcessType, BussProcess, AuthRolePriv, AuthUserPriv tables are filled in the same way (Fig. 1), and contain additional information about the authorized users of the system. Figure 1 shows the relationship of the tables in the User block.

3.2   Information on Measurements of the Concentration of Impurities of Specific Heavy Metals in the Atmosphere

The main table in this database is a table with data on measurements of the concentration of impurities of specific heavy metals in the atmosphere - Pollution, the types of materials and their physical characteristics, which may have implications for collection and analysis. Figure 2 shows the "Pollution" block.

This database in the future will consist of two parts. One of them corresponds to standard mathematical models of media, the other to models created in the natural experiment. In the course of the work, this database should be expanded and supplemented with new data.

Below describe the structure of the created tables (Tables 6, 7 and 8).

Table 6. DataCollectionPoint – basic information about data collection points

Fields	Data type	Field description	Field properties
id	bigint	Data collection point identifier	Key field
name	Varchar	Data collection point name	Obligatory field
lon	float	Longitude	Obligatory field
lat	float	Latitude	Obligatory field
cmt_ru	text	Comments	optional field

Data on heavy metals and other pollutants in the atmospheric air are stored in the PollutionElements table.

Table 7. PollutionElements

Fields	Data type	Field description	Field properties
Id	Bigint	Records ID	Key field of the table
name	varchar	Name of pollutant	Required, foreign key
alias	varchar	Secondary name (formula)	Optional
concentration	Float	Concentration of the contaminant in the atmosphere	Required, foreign key

The Measurements table contains data on measurements at collection points.

3.3   Wind Information

The wind is the key factor in the spread of pollutants in the ambient air, and it is advisable to create a separate data block for the wind parameters. The FD_WindType, Wind (Tables 9 and 10) contains basic information about the characteristics of the wind.

All tables can be added to the database as needed. Five tables are designed to write system data – Dblog, dbFileStorage, sys_jc_GenId, DbInfo, Wax_VerDb ModelingOthers, which will provide information about the files, the database.

To collect and process data to optimize the monitoring of air pollution, a database is created, which is shown in Fig. 3 as a relational model of the database.

Table 8. Measurements

Fields	Data type	Field description	Field Properties
Id	bigint	Records ID	Key field
point	bigint	data collection points	Obligatory, Key
dBeg	datetime	Start time of the conducted measurement	Obligatory
dEnd	datetime	End time of the conducted measurement	Obligatory
element	bigint	Contaminant	Obligatory, Key
temperature	float	Air temperature during measurement	Obligatory
pressure	float	Air pressure during measurement	Obligatory
value	float	Value	Obligatory

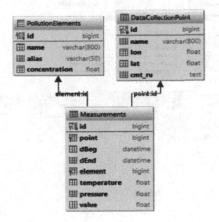

Fig. 2. Block "Pollution"

Table 9. FD_WindType

Fields	Data type	Field description	Field Properties
Id	bigint	Records ID	Key field
name_ru	varchar	Name	Obligatory, Key
fullName_ru	varchar	Full name	Obligatory
ord	bigint		values are allowed NULL
vis	bigint		values are allowed NULL

Table 10. Wind

Fields	Data type	Field description	Field Properties
Id	bigint	Records ID	Key field
dte	datetime	Measurement time	Obligatory
speed	float	Wind speed	Obligatory
windType	bigint	Type of wind	Obligatory, Key

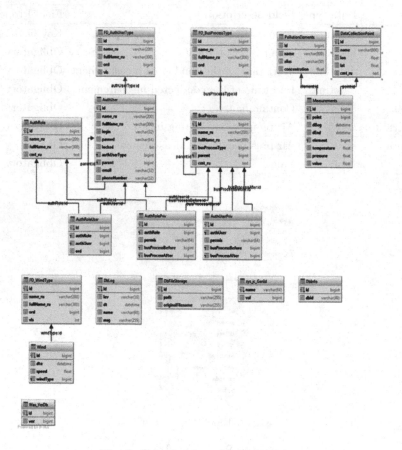

Fig. 3. Relational model of database

It is necessary to fix special marks on the surface that affect the measurements. After the work is finished and the data is viewed, the data is saved to a file that must be loaded into the database.

Next, we will develop an application that will transfer to the database the results of both mathematical modeling and experiments in accordance with the structure of the database.

4   General Structure of the Monitoring System

The working process of the air pollution monitoring and surveillance system is shown in Fig. 4 and consists of three main blocks, namely: a database, a real-time sensor data reporting system from two monitoring stations and the server. Finally, the results of the previous steps can be used for applications.

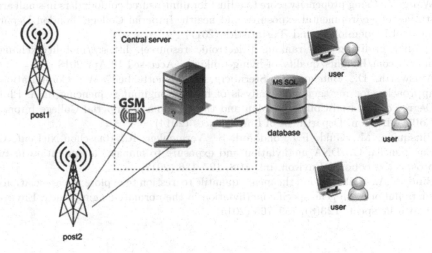

Fig. 4. General structure of the monitoring system

5   Conclusion

Despite all the efforts made by the Government of Almaty to improve the quality of atmospheric air in the city, air pollution remains the main environmental problem of our metropolis. The constant growth of the fleet of vehicles in combination with the growth of the number of industrial enterprises and the increase in the capacity of enterprises of the energy complex lead specialists to forecast a negative trend in the development of the ecological situation. In this connection, the use of modern technologies of spatial processing and data analysis makes it possible to significantly improve the efficiency of air quality management. The integrated approach implemented in the information system allows combining the maintenance of databases of emission sources, instrumental and calculation methods for air quality control, as well as using spatial analysis tools in organizing information support for making managerial decisions of the urban planning. The adoption of environmentally friendly management decisions is the basis for sustainable development of the city and the region as a whole.

The work was supported by the Ministry of Education and Science of Republic of Kazakhstan, under the Grant Development of a new information system and database to optimize monitoring of atmospheric air pollution by heavy metals (contract number 132, 12 March, 2018).

References

1. Zhumagulov, B.T., Temirbekov, N.M., Turganbaev, E.M., Denisova, N.F., Bitimbayev, I.T.: MInformation system of ecological monitoring and application of variational algorithms. Bull. Nat. Eng. Acad. Repub. Kaz. **1**, 10–18 (2013)
2. Rakhmutellina, S.Zh.: The forecasting subsystem of the air pollution monitoring information system. Search **2**, 243–250 (2010)
3. Wang, Y.: Using propensity score to adjust for unmeasured confounders in small area studies of environmental exposures and health. Imperial College. London Department of Epidemiology and Biostatistics (2017)
4. Quality of living city ranking [Electronic resource]. https://mobilityexchange. mercer.com/Insights/quality-of-living-rankings. Accessed 10 Apr 2018
5. Ainsworth, D., Butcher, S., Sternberg, M., Knottenbelt, W.: Computational approaches for metagenomic analysis of high-throughput sequencing data. Ph.D. Degree of Imperial College London and the Diploma of Imperial College Imperial College London, Department of Life Sciences (2014)
6. Plusquina, M., Guid, F., Polidorod, S., Vermeulen, R., Raaschou-Nielsenf, O., Campanella, G.: DNA methylation and exposure to ambient air pollution in two prospective cohorts. Environ. Int. **108**, 127–136 (2017)
7. Bind, M.A., et al.: Beyond the mean: quantile regression to explore the association of air pollution with gene-specific methylation in the normative aging study. Environ. Health Perspect. **123**(8), 759–765 (2015)

Temperature Field Reconstruction of Flame from Images for Optimal Energy-Efficient Control of the Air-Fuel Mixture Making in Steam – Driven Boilers

O. B. Ospanov[✉], D. L. Alontseva, and A. L. Krasavin

Department of Instrument Engineering and Technology Process Automation,
D. Serikbayev East Kazakhstan State Technical University,
Protozanov 69, 070004 Ust-Kamenogorsk, Kazakhstan
ospanovolzhas@gmail.com, dalontseva@mail.ru, alexanderkrasavin@mail.ru

Abstract. Over recent years, substantial efforts have been made to develop technologies for the flame 3D reconstruction and characterization. Among those, the digital image based on tomographic technique has received great attention due to its clear advantages over other approaches. The technology of digital image appeared to be the most suitable for the 3D measurement of flame in practical furnace. Digital image-based tomography of flame can be achieved using either a single camera or multi-camera set-up. Although the single camera approach is simple in structure and low in cost, it can only be applied under strict conditions where the flame is steady and has a high level of rotational symmetry. For the more accurate reconstruction of unsteady and asymmetric flame, a multi-camera system has to be employed. The process of capturing the light from a combustion flame onto an imaging sensor is physically equivalent to a Radon transformation where a 2D flame cross-section undergoes transformation to produce a 1D section projection. Consequently, the reconstruction of a flame cross-section from its multiple 1D projections turns out to be the inverse Radon transformation. Among the majority of currently known multi-camera optical pyrometry methods for temperature field reconstructing, the back-projection algorithm is used for Radon inverse transformation. At the same time, the back-projection algorithm is known to have a low accuracy of restoring the original distribution. The purpose of this report is a study of the algebraic approach possibilities to numerical Radon inverse transformation. We hope that the proposed method can be useful in the process of the 3D image temperature field reconstructing.

Keywords: Two-colorpyrometry · Flame ·
Temperature field reconstruction · Radon transform ·
Inverse Radon transform · Numerical methods

© Springer Nature Switzerland AG 2019
Y. Shokin and Z. Shaimardanov (Eds.): CITech 2018, CCIS 998, pp. 171–180, 2019.
https://doi.org/10.1007/978-3-030-12203-4_17

1 Introduction

In recent years, a large amount of research has been focused on the development of automatic process control systems in thermal power plants (CHP). The main control task for the CHP is the following: the automatic control system must regulate the output power to meet the electricity demand in combination with the safe maintenance of the required values of the main dynamic variables such as steam heater temperature, throttle pressure, furnace pressure, and drum water level. As a rule, this goal is achieved through the use of multi-level automatic controllers based on proportional-integral-differential (PID) controllers [1–3]. This approach turns out to be highly reliable and provides satisfactory performance during normal operation, maintained in the system, where the characteristics of CHP plants remain almost constant. However, the demand for electricity increases, while the magnitude of the cyclic change in the grid load of renewable sources such as wind, solar and hydropower, is strongly influenced by the season and weather conditions. Consequently, CHP must participate in the overall energy production network and respond to these changes by adjusting the load in a wide and rapidly changing wide range. Thus, the above thermal parameters of CHP plants should be monitored so that they can work optimally at any time. At present, the problems of control in conditions of non-linearity of parameters in a wide range of operations, high inertia and time-varying behavior, as well as strong interaction among a multitude of variables, are becoming serious challenges for CHP process control systems. Consequently, conventional PI/PID controllers based on controllers are already insufficient for the required performance, even if they are well tuned at a given load level. On the other hand, with the help of modern computer and instrumental technologies and the use of a distributed control system, it is now possible to introduce advanced intelligent controllers. Intelligent control systems developed on the basis of neuro-fuzzy networks are described in papers [4–6], where both linear models and controllers are given and evaluated. Thanks to simulation studies of a coordinated control system for a boiler-turbine system and a superheater controlled by an intelligent controller, higher control values are shown in comparison with linear intelligent systems of general control.

The authors of this work are developing an intelligent controller to control the processes of formation of the fuel-air mixture in steam boilers, including a "fast" temperature control circuit based on the readings of the pyrometer. A radiation detector is used as a detector, the selected detection wavelength depends on the type of fuel burned, the operation of the device and the circuits are described in more detail in our paper [7].

It has been proved that the geometric, luminous, and fluid dynamic characteristics of the flame in combustion systems are closely related to combustion efficiency and pollutant emissions as well as furnace safety [8]. Advanced monitoring and characterization of these flames have therefore become increasingly important for engineers to better understand and optimize combustion processes. In recent years, much research effort has been directed towards the development of advanced instrumentation systems for enhanced monitoring and specification of flames, particularly with the latest optical sensing, digital imaging, and image processing techniques [9]. However, the flame data obtained from these systems is limited to two dimensions (2-D) only, i.e., the third dimension has not been taken into account. The most recent advance in this area of research is the development of an instrumentation system for three-dimensional (3-D) visualization and quantitative characterization of red gas flames [9,10]. However, the system was elaborated to measure geometric and luminous parameters of flames from a reconstructed 3-D flame model, ignoring 3-D temperature information. Accurate measurement of temperature distribution in combustion flames is necessary to achieve fundamental understanding of combustion and polluting substances formation processes. Three-dimensional temperature measurement of combustion flames is of a major technical challenge. As a result, very limited scope of work has previously been reported on the subject. It is the nature of combustion flames which is the greatest problem in attempting to determine any information as to their internal structure in 3-D, including temperature. Moreover, as translucent, a flame cannot be inspected and analyzed using established 3-D techniques, such as tomography as used in medical imaging. Previous work in the area includes theoretical study on the determination of 3-D temperature distribution in a large furnace using multiple cameras [11]. However, the principles upon which the described system is based appear to be difficult for implementing the system.

The purpose of this study is to develop a mathematical apparatus for the reconstruction of the three-dimensional temperature field of the flame, which is being formed in an industrial boiler and controlled according to the readings of the pyrometer.

2   Principles of Two-Color Pyrometry

Pyrometry is based on the fact that all surfaces at temperatures above absolute zero emit thermal radiation. Planck's radiation law (1), modified to include surface emissivity, is the fundamental relation of thermal radiation:

$$M(\lambda, T) = \epsilon_\lambda \frac{C_1}{\lambda^s (e^{\frac{C_2}{\lambda T}} - 1)},\tag{1}$$

where $M(\lambda, T)$ is the monochromatic exitance, namely monochromatic radiation measured in ration of energy per unit area per unit time (W/m2/s), λ is the wavelength of the radiation (mm), T is absolute temperature (K), ϵ_λ is monochromatic emissivity, and C_1 and C_2 are the first and second Planck's constants. In cases $\frac{C_2}{\lambda T} \gg 1$, Planck's law can be replaced by the Wien's radiation law (2):

$$M(\lambda, T) = \epsilon_\lambda \frac{C_1}{(\lambda^s)^5} e^{\frac{C_2}{\lambda T}}.\tag{2}$$

It can be shown that the output of the imaging system, namely the grayscale $G(\lambda, T)$, is proportional to the exitance of the measured object and dependent on the spectral sensitivity S_λ of the imaging system (CCD sensor), i.e.,

$$G(\lambda, T) = RS_\lambda \epsilon_\lambda \frac{C_1}{(\lambda^s)^5} e \frac{C_2}{\lambda T}, \tag{3}$$

where R is called the instrument constant which is independent of wavelength and reflects the effects of various factors including radiation attenuation due to the optical system and atmosphere, observation distance, lens properties and signal conversion. The ratio between the grey levels at wavelengths λ_1 and λ_2 is given by Eq. (4):

$$\frac{G(\lambda_1, T)}{G(\lambda_2, T)} = \frac{S_{\lambda_1} \epsilon_{\lambda_1}}{S_{\lambda_2} \epsilon_{\lambda_2}} \left(\frac{\lambda_2}{\Lambda_1}\right)^5 \exp\left(\frac{C_2}{T}\left(\frac{1}{\lambda_2} - \frac{1}{\lambda_1}\right)\right) \tag{4}$$

The ratio between the spectral sensitivities $\frac{S_{(\lambda_2)}}{S_{(\lambda_1)}}$ is called instrument factor and is known from calibration using a tungsten lamp as a standard temperature source. A crucial point here is how to deal with the ratio between the spectral emissivities λ_1 and λ_2. Normally, gray body behaviour is assumed for the detected object (i.e., $\frac{\lambda_2}{\lambda_1} = 1$), when the wavelengths are very close to each other. For gaseous flames, the size of the soot particles ranges from 0.005 to 0.1 mm [5] and is much smaller than the wavelengths of observation. Flower (1983) proved that, when assuming the soot particles in the flame to be homogeneous, optically thin, isothermal along a horizontal line through the flame, and small relative to the used wavelength, the spectral emissivity is inversely proportional to the wavelength, i.e., $\lambda_1 \backsim \frac{1}{1}$. Therefore, another formulation for gaseous flames in this study is obtained by substituting $\frac{\lambda_2}{\lambda_1} = \frac{e(\lambda_2)}{e(\lambda_1)}$ into Eq. (5):

$$T = \frac{C_2\left(\frac{1}{\lambda_1} - \frac{1}{\lambda_2}\right)}{\ln \frac{G(\lambda_1, T)}{G(\lambda_2, T)} + \ln \frac{S_{\lambda_2}}{S_{\lambda_1}} + \ln \left(\frac{\lambda_1}{\lambda_2}\right)^6} \tag{5}$$

3   Principle of Measurement

Two-color pyrometry has been used extensively in many combustion applications but its application is relatively new to coal flames. Reviews and discussions on the uncertainties of the two-color method for measuring temperature and KL can be found by Ladommatos and Zhao [12] and others [13–16]. the two colors can be derived by several methods. The simplest is to obtain emission from a single line of sight (provides a point measurement) which is split and then optically altered to produce two narrow bands of light. Shaw and Essenhigh [17] used this method on a laboratory scale reactor using pulverized coal to yield point temperature measurements. Lu and Yan [9] used a single, CCD camera to measure the two-dimensional (2D) temperature in a 500 kW pulverized-coal flame. The light from the flame was split and altered into three narrow wavelength beams and captured by the CCD detector on the camera. The signals for the three beams were processed continuously to provide online temperature readings. The effects on temperature with varied air-fuel ratios, fuel low rates, and particle sizes were examined. Huang et al. [18] used a similar method to yield 2D, continuous temperature measurements using a single CCD camera with rotating, narrow (10 nm) band-pass filters. Three hundred images were taken over a period of thirty seconds with each band-pass filter. The signals for each wavelength were averaged and then used to compute the average, 2D flame temperature. In order to calculate the temperature distribution of a flame cross section from its gray scale images using the two-color method, it is necessary to reconstruct two gray scale representations of the cross section for two different spectral bands (Fig. 1).

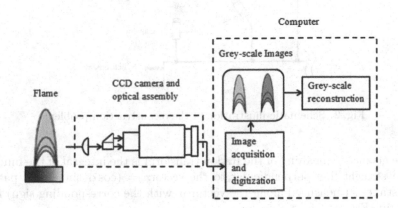

Fig. 1. Functional schematic of the two-color pyrometer

The process of projecting the light from a combustion flame onto an imaging sensor is physically equivalent to a Radon transformation, where a 2-D flame cross section undergoes transformation to produce a 1-D section projection [19]. Consequently, the reconstruction of a flame cross section from its 1-D projection is essentially the inverse Radon transformation. As a matter of fact, the implementation of the inverse Radon transformation requires infinitely many of projections from a 180 arc around the burner axis. In practice, however, the reconstruction is achieved using a sufficiently large number of projections (Fig. 2).

4   Radon Transform

Let $f(x, y)$ be a function of two real variables defined on the whole plane, and decreasing sufficiently fast at infinity (so that the corresponding improper integrals converge). Then by the Radon transform of the function $f(x, y)$ is defined as (6)

$$R(s, \alpha) = \int f(s \cos \alpha - z \sin \alpha, s \sin \alpha + z \cos \alpha) \, dz \qquad (6)$$

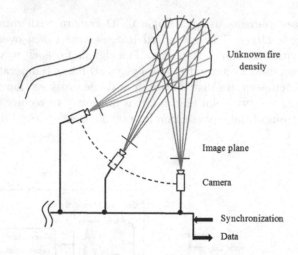

Fig. 2. Schematic illustration of the reconstruction problem

The geometric meaning of the Radon transform is the integral of the function along a straight line perpendicular to the vector $n = (\cos\alpha, \sin\alpha)$ and passing at a distance s (measured along the vector n, with the corresponding sign) from the origin (Fig. 3)

The Radon transform can be defined in different ways. The Radon transform $q^{\triangleq}(\rho, \tau)$ of a continuous two dimensional function $q(x, y)$ is found by stacking or integrating values of g along slanted lines. The location of the line is determined from the line parameter: slope ρ and line τ offset.

$$q^{\triangleq}(\rho, \tau) = \int q(x, \rho x + \tau)\, dx \qquad (7)$$

We will use

$$S(\rho, \varphi) = \iint f(x, y)\delta(\rho - x\cos\varphi + y\sin\varphi)\, dx dy, \qquad (8)$$

where $\delta(\rho)$ is a Dirac's delta-function.If the distribution function $f(x, y)$ is known, the formula (8) allows to calculate so-called "sinogram" $S(\rho, \varphi)$.

5   Numerical Inversion of Radon Transform

The inverse task means that to find $f(x, y)$ when the function $S(\rho, \varphi)$ is known. There is an exact solution of Eq. (1), which was found by Radon himself. It is called inverse Radon transform and has the form (9):

$$f(x, y) = -\frac{1}{2\pi^2} \int_0^\pi d\varphi \int \frac{DS}{D\rho} \frac{d\rho}{\rho - x\cos\varphi + y\sin\varphi} \qquad (9)$$

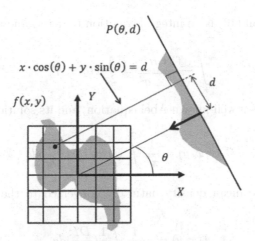

Fig. 3. Geometry of the Radon transforms

It should be noticed, that the method of Radon transform is widely used in tomography [20]. Its main advantage is that there is no need to solve the system of algebraic equations. This method gives the explicit expression for the density function $f(x, y)$ through the ray-sums of sinogram $S(\rho, \varphi)$, Of course the matter is not reduced to find the inverse operator for Eq. (12), which is incorrect. Currently, the most common method of numerical inversion of the Radon transform is the inverse projection algorithm [21, 22]. In any case we can get the corresponding expression by using different mathematical transformations and hypotheses. It seems that those transformations must be examined in detail and replaced by its approximated matrix analogs. In theory of Radon transform there is an interesting ratio, which gives an integral equation to find the phantom density. Let consider double integral on a plane (10):

$$I(q, x, y) = \iint_{D} \frac{f(\alpha, \beta)}{\sqrt{(x - \alpha)^2 + (x - \beta)^2 - q^2}} \, d\alpha d\beta, \qquad (10)$$

where D defined as $(x - \alpha)^2 + (x - \beta)^2 > r^2$

Let's

$$\alpha = x + r \cos \theta \beta = y + r \sin \theta \qquad (11)$$

Then (11) transforms to

$$\int_{r}^{\infty} \frac{r \, dr}{\sqrt{r^2 - q^2}} \int_{0}^{2\pi} (x + r \cos \theta, y + r \sin \theta) \, d\theta = I(q, x, y) \qquad (12)$$

The internal in (12) is a mean density along circles with radius r > q, that means

$$f^{\triangleq}(r, x, y) = \frac{1}{2\pi} \int_{0}^{2\pi} (x + r \cos \theta, y + r \sin \theta) \, d\theta \qquad (13)$$

Thus, expression (13) is an integral equation to determine unknown function $f(r, x, y)$:

$$\int_r^\infty \frac{f^{\triangle} r \, dr}{\sqrt{r^2 - q^2}} = \frac{1}{2\pi} I(q, x, y) \tag{14}$$

Last equation is a well-known Abel equation, and its solution is expressed by formula:

$$f^{\triangle}(r, x, y) = \frac{1}{\pi^2} \int_r^\infty \sqrt{r^2 - q^2} \left(\frac{DI}{Dq}\right) dq \tag{15}$$

For $r \to 1$: the mean density naturally converted to the density in point (x, y), and we get:

$$f(x, y) = -\frac{1}{\pi^2} \int_r^\infty \frac{1}{q} \left(\frac{DI}{Dq}\right) dq \tag{16}$$

If we know relation between distances q and detector numbers p, then the function $I(q, x, y)$ can be written through the sinogram in form

$$I(q, x, y) = \int_0^{2\pi} S(\rho, \varphi) \, d\varphi \tag{17}$$

Hence, the phantom density (17) can be presented as (18):

$$f(x, y) = \int_0^{2\pi} d\varphi \int \frac{DS}{D\rho} \frac{d\rho}{q(\rho, \varphi, x, y)} \tag{18}$$

The value q, included into (7) is a distance from each image point (x, y) to each ray (ρ, φ) passing camera. The elements of four dimensional array q can be calculated by formula (7). For its practical usage one can introduce three dimensional array of inversed distances $Q(\alpha, x, y)$. In formula (6), which made the Radon transform, exist a numerical operation of differentiation of sinogram $S(\rho, \varphi)$ by detectors ρ. If ρ considered as number of lines, and φ - as number of columns, then numerical differentiation is more comfortable do by multiplication of $S(\rho, \varphi)$ from left on the corresponding matrix M, i.e. $S_1(\rho, \varphi) = DS(\rho, \varphi) = MS(\rho, \varphi)$. In the internal points of differentiated vector-column the derivative is approximated by central differences, but in extreme points - one-side differences. The matrices of differentiation operator M can be built with any sizes. They are distinguished from Toeplitz matrices, having the same values on diagonals, by the first and last lines.

The main results of this research are following:

(1) The mathematical approach of the two-color pyrometry task with usage matrix algorithm for 3D temperature field reconstruction was formulated.

(2) In contrast to the known approaches, the matrix algorithm gives precise reconstruction of the radiation intensity distribution. However, it is sensitive to the inaccuracy of image parameters. This drawback is also inherent for other image based pyrometry methods.
(3) We have currently tested the possibility of using the direct method of solving the inverse Radon equation in combination with the recently developed regularization methods.

6   Conclusion

A mathematical apparatus has been developed to recreate the three-dimensional temperature field of the flame, which is being formed in an industrial boiler unit and controlled according to the readings of the pyrometer. The two-color pyrometry problem is solved using the matrix algorithm for the recovery of a three-dimensional temperature field using direct and inverse Radon transforms. In practice, the developed methods are used to create an intelligent system for managing the processes of forming a fuel-air mixture in steam boilers, including a "fast" temperature control circuit with an intelligent controller that works with pyrometer readings.

The results of the research are of significance for a wide range of researchers developing temperature control circuits using pyrometers.

References

1. Xiao, W., Jiong, S., Yiguo, L., Kwang, Y.L.: Steam power plant configuration, design, and control. In: WIREs Energy Environ, p. 27. Wiley, Hoboken (2015)
2. Mallik, A.: State feedback based control of air-fuel-ratio using two wide-band oxygen sensors. In: Proceedings of 10th Asian Control Conference (ASCC 2015), Kota Kinabalu, Malaysia, pp. 1–6 (2015)
3. Najimi, E., Ramezani, M.H.: Robust control of speed and temperature in a power plant gas turbine. ISA Trans. 51, 304–308 (2012)
4. Li, X., Guan, P., Chan, C.W.: Nonlinear multivariable power plant coordinate control by constrained predictive scheme. IEEE Trans. Contr. Sys. Technol. 18, 1116–1125 (2010)
5. Liu, X.J., Chan, C.W.: Neuro-fuzzy generalized predictive control of boiler steam temperature. IEEE Trans. Energy Conver. 21, 900–908 (2006)
6. Liu, X., Liu, J.: Constrained power plant coordinated predictive control using neurofuzzy model. Acta Autom. Sinica 32, 785–790 (2006)
7. Ospanov, O.B., Alontseva, D.L. Krasavin, A.L.: Development of an intelligent system for optimal energy-efficient control of the air-fuel mixture making in steam - driven boilers. In: The Joint Issue of Journals "Herald of D. Serikbayev EKSTU" and "Computational Technologies", vol. 1, no. 2, pp. 56–70 (2018)
8. Toftegaard, M.J., Brix, J., Jensen, P.A., Glarborg, P., Jensen, A.D.: Oxy-fuel combustion of solid fuels. Prog. Energy Combust. Sci. 36(5), 581–625 (2010)
9. Lu, G., Yan, Y.: Temperature profiling of pulverized coal flame using multicolor pyrometric and digital imaging techniques. IEEE Trans. Instrum. Meas. 55(4), 1303–1308 (2006)

10. Brisley, P.B., Lu, G., Yan, Y., Cornwell, S.: Three dimensional temperature measurement of combustion flames using a single monochromatic CCD camera. IEEE Trans. Instrum. Meas. **54**(4), 1417–1421 (2005)
11. Zhou, H.C., Han, S.D., Lou, C., Liu, H.: A new model of radiative image formation used in visualization of 3-D temperature distributions in large-scale furnaces. Numer. Heat Transfer B Fundam. **42**(3), 243–258 (2002)
12. Ladommatos, N., Zhao, H.: A guide to measurement of flame temperature and soot concentration in diesel engines using the two-colour method, part 1. In: SAE Paper, vol. 941956 (1994)
13. Ladommatos, N., Zhao, H.: A guide to measurement of flame temperature and soot concentration in diesel engines using the two-colour method, part 2. In: SAE Paper, vol. 941956 (1994)
14. di Stasio, S., Massoli, P.: Influence of the soot property uncertainties in temperature and volume-fraction measurements by two-colour pyrometry. Meas. Sci. Technol. **5**, 1453–1465 (1994)
15. Zhao, H., Ladommatos, N.: Optical diagnostics for soot and temperature measurement in diesel engines. Prog. Energy Combust. Sci. **24**(3), 221–225 (1998)
16. Tree, D.R., Svensson, K.I.: Soot processes in compression ignition engines. Prog. Energy Combust. Sci. **87**, 272–309 (2007)
17. Shaw, D.W., Essenhigh, R.H.: Temperature fluctuations in pulverized coal (P.C.) flames. Combust. Flame **86**, 333–346 (1991)
18. Huang, Y., Yan, Y., Riley, G.: Vision-based measurement of temperature distribution in a 500-kW model furnace using the two-colour method. Measurement **28**, 175–183 (2000)
19. Lou, C., Zhou, H.C., Yu, P.F., Jiang, Z.W.: Measurements of the flame emissivity and radiative properties of particulate medium in pulverized-coal-fired boiler furnaces by image processing of visible radiation. Proc. Combust. Inst. **31**(2), 2771–2778 (2007)
20. Faridani, A.: Tomography and sampling theory, the radon transform and applications to inverse problems. In: AMS Proceedings of Symposia in Applied Mathematics. American Mathematical Society, Providence (2006)
21. Jia, R.X., Xiong, Q.Y., Wang, K., Wang, L.J., Xu, G.Y., Liang, S.: The study of three-dimensional temperature field distribution reconstruction using ultrasonic thermometry. AIP Adv. **6**(7), 075007 (2016)
22. Li, Y., Liu, S., Inaki, S.H.: Dynamic reconstruction algorithm of three-dimensional temperature field measurement by acoustic tomography. Sensors **17**(9), 2084 (2017)

Quasi-geostrophic Wave Motion in a Rotating Layer of Electrically Conducting Fluid with Consideration of Dissipation Effects

S. I. Peregudin[1]($\boxtimes$), E. S. Peregudina[2], and S. E. Kholodova[3]

[1] St. Petersburg State University, Saint Petersburg, Russia
peregudinsi@yandex.ru
[2] St. Petersburg Mining University, Saint Petersburg, Russia
ehllina-peregudina@yandex.ru
[3] ITMO University, Saint Petersburg, Russia
kholodovase@yandex.ru

Abstract. The purpose of the paper is to obtain an analytic solution of the boundary-value problem for the system of nonlinear partial differential equations that model magnetohydrodynamic perturbations in a layer of perfect electrically conducting rotating fluid bounded by space- and time-varying surfaces with due account of the dissipative factors of the magnetic field diffusion, the inertia forces, and the Coriolis force.

We construct an exact solution of the reduced nonlinear equations that describes the propagation of waves of finite amplitude in an infinite horizontally extended electrically conducting fluid when the surface bounding the layer has approximatively constant gradient over distances of the order of the wavelength.

Keywords: Quasi-geostrophic motion · Rotating fluid · Magnetic field diffusion · Magnetohydrodynamic processes

1 Introduction

The paper is concerned with theoretical studies of nonlinear wave processes in liquid electrically conducting media with dissipation which are influenced by magnetic forces. It is known that a motion of conducting media in a magnetic field is accompanied by interaction effects between mechanical and electromagnetic fields due to ponderomotive forces and elastic motions. Various manifestations of such interaction is the subject of study of magnetic hydrodynamics, one of the branches of continuum mechanics, which was considerably advanced by H. Alfvén, G. G. Branover, T. G. Cowling, A. G. Kulikovskii, G. A. Lyubimov, J. A. Shercliff, and other researchers, who studied propagation of small perturbations in ideal media and succeeded in solving hydrodynamically classical motion problems of conducting fluids in channels and tubes.

© Springer Nature Switzerland AG 2019
Y. Shokin and Z. Shaimardanov (Eds.): CITech 2018, CCIS 998, pp. 181–188, 2019.
https://doi.org/10.1007/978-3-030-12203-4_18

Practical considerations indicated the need in the development of simplified mathematical models of MHD-processes more suitable for dealing with specific problems. In this direction, one should mention the models of weakly conducting media obtained by S. I. Braginskii, G. G. Branover, A. G. Kulikovskii, and G. A. Lyubimov, as well as the asymptotically degenerated models of magnetic hydrodynamics with small and large Reynolds numbers.

Within the framework of the above approximations it proved possible to construct the solution of many practically important problems of flow of conducting media and investigate processes of propagation of linear magnetohydrodynamic waves both in homogeneous and in inhomogeneous media.

The study of nonlinear MHD waves started relatively recently. Principal results in this problem were obtained by F. De Hoffmann in the approximation of weak conductivity of the fluid without consideration of rotation and by K. D. Danov and M. S. Ruderman for a perfectly conducting media.

In the present study we construct a nonlinear mathematical model of dynamic wave processes in an electrically conducting fluid with due account of such important effects as the rotation of the medium and dissipation associated with equiprobable effect of the convective and diffusion terms in the magnetic field induction equation.

2   Quasi-geostrophic Motion. Statement and Solution of the Problem

Let us consider a rotating layer of perfect incompressible electrically conducting fluid bounded from above by a hard impermeable surface $z = Z(x, y)$, and from below, by the surface $z = -h_B(x, y, t)$. We use the rectangular Cartesian coordinates $Oxyz$. By the body force we shall mean the vector $\mathbf{g}$ orthogonal to the surface $z = 0$ and antiparallel to the vertical axis. The rotation axis of the fluid coincides with the z-axis; i.e., $\boldsymbol{\omega} = \mathbf{k}\omega$, where $\boldsymbol{\omega}$ is the angular velocity of rotation of the layer. We solve the problem in the framework of long wavelength approximation [1],

$$\frac{\partial v_x}{\partial t} + v_x \frac{\partial v_x}{\partial x} + v_y \frac{\partial v_x}{\partial y} = 2\omega v_y + g\frac{\partial \eta}{\partial x} + \frac{1}{\mu\rho}\left(b_x \frac{\partial b_x}{\partial x} + b_y \frac{\partial b_x}{\partial y}\right),$$

$$\frac{\partial v_y}{\partial t} + v_x \frac{\partial v_y}{\partial x} + v_y \frac{\partial v_y}{\partial y} = -2\omega v_x + g\frac{\partial \eta}{\partial y} + \frac{1}{\mu\rho}\left(b_x \frac{\partial b_y}{\partial x} + b_y \frac{\partial b_y}{\partial y}\right),$$

$$\frac{\partial \eta}{\partial t} + \frac{\partial}{\partial x}\left[(H_0 + \eta)v_x\right] + \frac{\partial}{\partial y}\left[(H_0 + \eta)v_y\right] = 0,$$

$$(h_B - Z)\left(\frac{\partial b_x}{\partial x} + \frac{\partial b_y}{\partial y}\right) + b_{z0}^{(e)}(x, y, t) - b_{z0}(x, y, t) = 0,$$

$$\frac{\partial b_x}{\partial t} + v_x \frac{\partial b_x}{\partial x} + v_y \frac{\partial b_x}{\partial y} - b_x \frac{\partial v_x}{\partial x} - b_y \frac{\partial v_x}{\partial y} = \lambda\left(\frac{\partial^2 b_x}{\partial x^2} + \frac{\partial^2 b_x}{\partial y^2}\right),$$

$$\frac{\partial b_y}{\partial t} + v_x \frac{\partial b_y}{\partial x} + v_y \frac{\partial b_y}{\partial y} - b_x \frac{\partial v_y}{\partial x} - b_y \frac{\partial v_y}{\partial y} = \lambda\left(\frac{\partial^2 b_y}{\partial x^2} + \frac{\partial^2 b_y}{\partial y^2}\right),$$

where $\lambda = \dfrac{1}{\sigma\mu}$ is the magnetic diffusion coefficient. Let L be the linear scale, U is the velocity scale, B is the scale of the magnetic field, T is the time scale, and N is the ordinate scale of the surface $\eta(x, y, t)$. Consider the dimensionless variables

$$x', \quad y', \quad t', \quad v'_x, \quad v'_y, \quad b'_x, \quad b'_y, \quad \eta', \quad x = Lx', \quad y = Ly',$$
$$t = Lt', \quad v_x = Uv'_x, \quad v_y = Uv'_y, \quad b_x = Bv'_x, \quad b_y = Bb'_y, \quad \eta = N\eta'.$$

Now the initial equations assume the form (dropping prime notation)

$$\varepsilon_T \frac{\partial v_x}{\partial t} + \varepsilon \left(v_x \frac{\partial v_x}{\partial x} + v_y \frac{\partial v_y}{\partial y} \right) - v_y = \frac{gN}{LaU} \frac{\partial \eta}{\partial x} + \frac{1}{\mu\rho} \left(\frac{B}{U} \right)^2 \varepsilon \left(b_x \frac{\partial b_x}{\partial x} + b_y \frac{\partial b_x}{\partial y} \right),$$

$$\varepsilon_T \frac{\partial v_y}{\partial t} + \varepsilon \left(v_x \frac{\partial v_y}{\partial x} + v_y \frac{\partial v_y}{\partial y} \right) + v_x = \frac{gN}{LaU} \frac{\partial \eta}{\partial y} + \frac{1}{\mu\rho} \left(\frac{B}{U} \right)^2 \varepsilon \left(b_x \frac{\partial b_y}{\partial x} + b_y \frac{\partial b_y}{\partial y} \right),$$

$$\varepsilon_T F \frac{\partial \eta}{\partial t} + \varepsilon F \left(v_x \frac{\partial \eta}{\partial x} + v_y \frac{\partial \eta}{\partial y} \right) - v_x \frac{\partial}{\partial x} \left(\frac{Z}{D} \right) - v_y \frac{\partial}{\partial y} \left(\frac{Z}{D} \right)$$
$$+ \left(1 + \varepsilon F \eta - \frac{Z}{D} \right) \left(\frac{\partial v_x}{\partial x} + \frac{\partial v_y}{\partial y} \right) = 0,$$

$$\left(1 + \varepsilon F \eta - \frac{Z}{D} \right) \left(\frac{\partial b_x}{\partial x} + \frac{\partial b_y}{\partial y} \right) + b_{z_0}^{(e)}(x, y, t) - b_{z_0}(x, y, t) = 0,$$

$$\varepsilon_T \frac{\partial b_x}{\partial t} + \varepsilon \left(v_x \frac{\partial b_x}{\partial x} + v_y \frac{\partial b_x}{\partial y} - b_x \frac{\partial v_x}{\partial x} - b_y \frac{\partial v_x}{\partial y} \right) = \frac{\varepsilon}{R_m} \left(\frac{\partial^2 b_x}{\partial x^2} + \frac{\partial^2 b_x}{\partial y^2} \right),$$

$$\varepsilon_T \frac{\partial b_y}{\partial t} + \varepsilon \left(v_x \frac{\partial b_y}{\partial x} + v_y \frac{\partial b_y}{\partial y} - b_x \frac{\partial v_y}{\partial x} - b_y \frac{\partial v_y}{\partial y} \right) = \frac{\varepsilon}{R_m} \left(\frac{\partial^2 b_y}{\partial x^2} + \frac{\partial^2 b_y}{\partial y^2} \right).$$

Here

$$\varepsilon_T = \frac{1}{T\alpha}, \quad \varepsilon = \frac{U}{L\alpha}, \quad F = \frac{\alpha^2 L^2}{gD}, \quad \alpha = 2\omega, \quad R_m = \frac{UL}{\lambda},$$

$H = H_0(x, y) + \eta(x, y, t) = D - Z + \eta$, H_0 is the unperturbed fluid depth and D is a constant.

We next assume that

$$\frac{gN}{LaU} = 1. \tag{1}$$

Hence $N = \dfrac{LaU}{g}$. From the above analysis it follows that $\dfrac{B^2}{\mu\rho U^2} = O(1)$, $F = O(1)$. Additionally, we assume that

$$\frac{Z}{D} = \varepsilon\eta_b, \qquad b_{z_0}^{(e)} - b_{z_0} = \varepsilon b_h, \tag{2}$$

where η_b, b_h are of the order of unity. Condition (1) means that if the Rossby number ε is small, then it is not yet so great that the motion differs substantially from a strictly geostrophic motion.

The Rossby numbers ε_T and ε measure the ratio of the local and advective accelerations to the Coriolis acceleration. The ratio of the local acceleration to the advective acceleration is controlled by the parameter

$$\frac{\varepsilon_T}{\varepsilon} = \frac{L}{UT}.$$

In the case when this parameter is large, the equations are essentially linear; i.e., the local time derivative is dominant over the nonlinear advective terms. Let us assume that the nonlinear terms are of the same importance as the local acceleration. In other words, we assume that

$$\frac{\varepsilon_T}{\varepsilon} = 1;$$

i.e., we consider the cases when the advection time $\dfrac{L}{U}$ has the same order as the time scale of local variations.

Using conditions (1), (2) and applying the operator rot to the first and second equations, we obtain, from the third equation of the above system for $\varepsilon_T = \varepsilon$ in the first approximation ($\varepsilon = 0$),

$$\left(\frac{\partial}{\partial t} + v_x \frac{\partial}{\partial x} + v_y \frac{\partial}{\partial y}\right)(\Omega - F\eta + \eta_b) = M\left(b_x \frac{\partial \zeta}{\partial x} + b_y \frac{\partial \zeta}{\partial y}\right);$$

$$v_x = \frac{\partial \eta}{\partial y}, \quad v_y = -\frac{\partial \eta}{\partial x}, \quad \Omega = -\Delta\eta, \quad \zeta = \frac{\partial b_y}{\partial x} - \frac{\partial b_x}{\partial y};$$

$$\frac{\partial b_x}{\partial x} + \frac{\partial b_y}{\partial y} = 0; \tag{3}$$

$$\frac{\partial b_x}{\partial t} + v_x \frac{\partial b_x}{\partial x} + v_y \frac{\partial b_x}{\partial y} - b_x \frac{\partial v_x}{\partial x} - b_y \frac{\partial v_x}{\partial y} = \frac{1}{R_m}\left(\frac{\partial^2 b_x}{\partial x^2} + \frac{\partial^2 b_x}{\partial y^2}\right),$$

$$\frac{\partial b_y}{\partial t} + v_x \frac{\partial b_y}{\partial x} + v_y \frac{\partial b_y}{\partial y} - b_x \frac{\partial v_y}{\partial x} - b_y \frac{\partial v_y}{\partial y} = \frac{1}{R_m}\left(\frac{\partial^2 b_y}{\partial x^2} + \frac{\partial^2 b_y}{\partial y^2}\right),$$

where $M = \dfrac{B^2}{\mu\rho U^2}$. In the absence of magnetic field, such approximation is called *quasi-geostrophic* in hydrodynamics; this means that the Rossby number is small, but at the same time it is still large in order that the fluid would move as a family of columns.

The solution to the system of nonlinear equations (3) will be sought as a functional dependence of components of the magnetic field on the function η,

$$b_x = f_1(\eta), \qquad b_y = f_2(\eta).$$

Now Eq. (3) in terms of the function η assume the form

$$\left(\frac{\partial}{\partial t} + \frac{\partial \eta}{\partial y}\frac{\partial}{\partial x} - \frac{\partial \eta}{\partial x}\frac{\partial}{\partial y}\right)(-\Delta\eta - F\eta + \eta_b) =$$

$$M\left[f_2''\left(f_1\left(\frac{\partial \eta}{\partial x}\right)^2 + f_2\frac{\partial \eta}{\partial x}\frac{\partial \eta}{\partial y}\right) - f_1''\left(f_1\frac{\partial \eta}{\partial x}\frac{\partial \eta}{\partial y} + f_2\left(\frac{\partial \eta}{\partial x}\right)^2\right)\right.$$

$$\left. + f_2'\left(f_1\frac{\partial^2\eta}{\partial x^2} + f_2\frac{\partial^2\eta}{\partial x\partial y}\right) - f_1'\left(f_1\frac{\partial^2\eta}{\partial x\partial y} + f_2\frac{\partial^2\eta}{\partial y^2}\right)\right], \tag{4}$$

$$f_1'\frac{\partial \eta}{\partial t} - f_1\frac{\partial^2\eta}{\partial x\partial y} - f_2\frac{\partial^2\eta}{\partial y^2} = \frac{1}{R_m}f_1''\left(\left(\frac{\partial \eta}{\partial x}\right)^2 + \left(\frac{\partial \eta}{\partial y}\right)^2\right) + \frac{1}{R_m}f_1'\Delta\eta, \tag{5}$$

$$f_2'\frac{\partial \eta}{\partial t} + f_1\frac{\partial^2\eta}{\partial x^2} + f_2\frac{\partial^2\eta}{\partial x\partial y} = \frac{1}{R_m}f_2''\left(\left(\frac{\partial \eta}{\partial x}\right)^2 + \left(\frac{\partial \eta}{\partial y}\right)^2\right) + \frac{1}{R_m}f_2'\Delta\eta. \tag{6}$$

The terms in the expression $-\Delta\eta - F\eta + \eta_b$ are completely determined by the relative motion. The term is a relative vorticity, the second term describes the contribution from the layer depth changes. The last term corresponds to the contribution from the surface topography $z = -Z(x, y)$ and is independent of the motion. The right-hand side of Eq. (4) contributes to the vorticity equation due to the presence of the magnetic field.

Thus, the problem of determining the quasi-geostrophic motion is reduced to the solution of the system of three nonlinear equations for perturbation of the surface η (or, what is the same, for the hydromagnetic pressure) and for the functions $f_1(\eta)$ and $f_2(\eta)$ that describe the magnetic field. From the available solutions η, f_1 and f_2 of Eqs. (4)–(6), one can find from the above relations the velocity components v_x, v_y and the field components b_x, b_y.

The solution η will be sought in the form

$$\eta = Ae^{i(kx + ly - \sigma t)}.$$

Now system (4)–(6) assumes the form

$$i\left(-\sigma(k^2 + l^2) + \sigma F + l\frac{\partial \eta_b}{\partial x} - k\frac{\partial \eta_b}{\partial y}\right) = M\left(f_1''(klf_1 + l^2 f_2)\right.$$

$$-f_2''(k^2 f_1 - klf_2))\,Ae^{i(kx + ly - \sigma t)} + \left(f_1'(klf_1 + l^2 f_2) - f_2'(k^2 f_1 + klf_2)\right), \tag{7}$$

$$-i\sigma f_1' + klf_1 + l^2 f_2 = -\frac{1}{R_m}f_1''(k^2 + l^2)Ae^{i(kx + ly - \sigma t)} - \frac{1}{R_m}f_1'(k^2 + l^2), \tag{8}$$

$$i\sigma f_2' + k^2 f_1 + klf_2 = \frac{1}{R_m}f_2''(k^2 + l^2)Ae^{i(kx + ly - \sigma t)} + \frac{1}{R_m}f_2'(k^2 + l^2). \tag{9}$$

Analytic expressions of the functions $f_1(\eta)$ and $f_2(\eta)$ can be found from the system of Eqs. (8), (9). Excluding the function f_2 from this system, we find that

$$f_2 = \frac{1}{l^2}\left(i\sigma f_1' - klf_1 - f_1''\frac{1}{R_m}(k^2 + l^2)\eta - f_1'\frac{1}{R_m}(k^2 + l^2)\right). \tag{10}$$

As a result, we have the equation

$$\frac{1}{R_m}(k^2 + l^2)\eta^2 f_1^{(4)} + 2(k^2 + l^2)\left(-i\sigma + \frac{2}{R_m}(k^2 + l^2)\right)\eta^2 f_1'''$$
$$+ \left(i\sigma - \frac{1}{R_m}(k^2 + l^2)\right)\left(i\sigma R_m - 2(k^2 + l^2)\right)f_1'' = 0$$

for the function $f_1(\eta)$. Its general solution reads as

$$f_1(\eta) = \frac{C_3}{(\lambda_1 + 1)(\lambda_1 + 2)}\eta^{\lambda_1 + 2} + \frac{C_4}{(\lambda_2 + 1)(\lambda_2 + 2)}\eta^{\lambda_2 + 2} + C_1\eta + C_2,$$

where

$$\lambda_{1,2} = \frac{a - b \pm \sqrt{(b - a)^2 - 4ac}}{2a}, \qquad a = \frac{k^2 + l^2}{R_m^2},$$

$$b = 2(k^2 + l^2)\left(-i\sigma + 2\frac{k^2 + l^2}{R_m}\right),$$

$$c = \left(-i\sigma - \frac{k^2 + l^2}{R_m^2}\right)\left(i\sigma R_m - 2(k^2 + l^2)\right).$$

The form of the function f_2 is determined by formula (10). But Eq. (7) is satisfied only by the functions f_1 and f_2 for which C_3 and C_4 are zero. Hence

$$f_1(\eta) = C_1\eta + C_2, \tag{11}$$

$$f_2(\eta) = \frac{i\sigma - \dfrac{k^2 + l^2}{R_m}}{l^2}C_1 - \frac{kC_2}{l} - \frac{kC_1}{l}\eta. \tag{12}$$

Using expression (11) and (12), Eq. (7) assumes the form

$$-\sigma(k^2 + l^2) + \sigma F + l\frac{\partial \eta_B}{\partial x} - k\frac{\partial \eta_B}{\partial y} = \frac{MC_1^2}{l^2}(k^2 + l^2)\left(\sigma + \frac{i(k^2 + l^2)}{R_m}\right). \tag{13}$$

Note that the equation representing the *quasi-solenoidality* of the magnetic field is satisfied identically.

We thus have the following conclusion. For an electrically conducting rotating fluid which is infinitely extended in the horizontal direction with $\nabla \eta_b = \text{const}$ (which is equivalent to the assumption that the slope of the surface $z = -Z(x, y)$ is nearly constant over distances of the order of the wavelength), the precise solution of the system of nonlinear equations (3) reads as

$$\eta = Ae^{i(kx + ly - \sigma t)}, \qquad v_x = \frac{\partial \eta}{\partial y}, \qquad v_y = -\frac{\partial \eta}{\partial x},$$

$$b_x = C_1\eta + C_2, \qquad b_y = \frac{i\sigma - \dfrac{kC_2}{R_m}}{l^2}C_1 - \frac{kC_2}{l} - \frac{kC_1}{l}\eta.$$

From Eq. (13) we have the following dispersion relation

$$\sigma = \frac{k\dfrac{\partial \eta_b}{\partial y} - l\dfrac{\partial \eta_b}{\partial x} + \dfrac{i(k^2 + l^2)C_1^2 M}{R_m l^2}}{F - \left(1 + \dfrac{k^2}{l^2}\right)(l^2 + C_1^2 M)}. \tag{14}$$

Note that in the case $C_1 M = 0$ the dispersion relation has the same form as for the low-frequency Rossby wave in a nonconducting fluid. In both cases, waves with higher frequency become filtered due to the a priori assumption on the quasi-geostrophic character of motion. In the limit $R_m \to \infty$ the dispersion relation (14) assumes the form of the dispersion relation obtained earlier.

In the real form, the principal motion characteristics read as

$$\eta(x, y, t) = A e^{\sigma_2 t} \cos(kx + ly - \sigma_1 t),$$

$$b_x(x, y, t) = (a\cos(kx + ly - \sigma_1 t) - b\sin(kx + ly - \sigma_1 t))\, A e^{\sigma_2 t} + C_2,$$

$$b_y(x, y, t) = \frac{k}{l}\,(b\sin(kx + ly - \sigma_1 t) - a\cos(kx + ly - \sigma_1 t))\, A e^{\sigma_2 t}$$

$$- \left(\frac{b\sigma_1 + a\sigma_2 + \frac{k^2+l^2}{R_m}a}{l^2} - \frac{kC_2}{l}\right),$$

$$v_x = -lA e^{\sigma_2 t} \sin(kx + ly - \sigma_1 t), \qquad v_y = kA e^{\sigma_2 t}\sin(kx + ly - \sigma_1 t).$$

Here

$$\sigma_1 = \frac{\left(k\dfrac{\partial \eta_b}{\partial y} - l\dfrac{\partial \eta_b}{\partial x} - \dfrac{2abM(k^2 + l^2)^2}{l^2 R_m}\right)\left(F - \left(1 + \dfrac{k^2}{l^2}\right)(l^2 + (a^2 - b^2)M)\right)}{\left(F - \left(1 + \dfrac{k^2}{l^2}\right)(l^2 + (a^2 - b^2)M)\right)^2 + 4a^2 b^2 M^2 \left(1 + \dfrac{k^2}{l^2}\right)^2}$$

$$- \frac{\dfrac{2abM^2(k^2 + l^2)^3(a^2 - b^2)}{l^4 R_m}}{\left(F - \left(1 + \dfrac{k^2}{l^2}\right)(l^2 + (a^2 - b^2)M)\right)^2 + 4a^2 b^2 M^2 \left(1 + \dfrac{k^2}{l^2}\right)^2},$$

$$\sigma_2 = \frac{2abM\left(1 + \dfrac{k^2}{l^2}\right)\left(k\dfrac{\partial \eta_b}{\partial y} - l\dfrac{\partial \eta_b}{\partial x} - \dfrac{2abM(k^2 + l^2)^2}{l^2 R_m}\right)}{\left(F - \left(1 + \dfrac{k^2}{l^2}\right)(l^2 + (a^2 - b^2))M\right)^2 + 4a^2 b^2 M^2 \left(1 + \dfrac{k^2}{l^2}\right)^2}$$

$$+ \frac{\dfrac{(k^2 + l^2)^2(a^2 - b^2)M}{l^2 R_m}\left(F - \left(1 + \dfrac{k^2}{l^2}\right)(l^2 + (a^2 - b^2)M)\right)}{\left(F - \left(1 + \dfrac{k^2}{l^2}\right)(l^2 + (a^2 - b^2))M\right)^2 + 4a^2 b^2 M^2 \left(1 + \dfrac{k^2}{l^2}\right)^2}.$$

where $a, b, \sigma_1, \sigma_2 \in \mathbb{R}, C_1 = a + ib, \sigma = \sigma_1 + i\sigma_2$. As $R_m \to \infty$ the sign of σ_2 depends on the sign on the expression $ab\left(k\dfrac{\partial \eta_b}{\partial y} - l\dfrac{\partial \eta_b}{\partial x}\right)$. For the existence of a bounded solution it is necessary to satisfy the inequality

$$ab\left(k\frac{\partial \eta_b}{\partial y} - l\frac{\partial \eta_b}{\partial x}\right) < 0.$$

By considering the projection of the magnetic induction equation to the vertical axis one can find a link between the amplitude of the external magnetic field, the topography of the layer outer surface, the oscillation amplitude of the interior boundary, and the amplitude of the generated magnetic field inside the liquid layer. Indeed, under the conditions of the problem under consideration, with $b_z = b_{z_0}^{(e)}$ the above equation assumes the form

$$\frac{\partial b_{z_0}^{(e)}}{\partial t} + \frac{\mathcal{D}(\eta, b_{z_0}^{(e)})}{\mathcal{D}(x, y)} + \frac{i\sigma - \dfrac{k^2 + l^2}{R_m}}{l^2} C_1 \frac{\mathcal{D}\left(\dfrac{\partial \eta}{\partial y}, Z\right)}{\mathcal{D}(x, y)} - \frac{1}{R_m}\Delta b_{z_0}^{(e)} = 0.$$

From this equation, for the periodic field $b_{z_0}^{(e)}$ with respect to the horizontal coordinates and time (i.e., for example, for $b_{z_0}^{(e)} = Be^{i(kx + ly - \sigma t)}$), it follows that

$$B = AC_1\left(\frac{k}{l}\frac{\partial Z}{\partial y} - \frac{\partial Z}{\partial x}\right). \tag{15}$$

It is worth noting that the above relation (15) has the same form as in the case of a frozen magnetic field.

From the above analytic solutions enable one can assess the effect of the topography of the domain dynamics on the magnetohydrodynamic characteristics of a wave process inside the liquid layer. The results obtained can be employed in astrophysics and geophysics. The layer boundaries are known be of great importance in its evolution, as well as in the dynamical processes occurring inside the liquid layer (for example, in geophysics the mantle topography controls the dynamic velocity of the solid core, which in turn can have an effect on the growth rate of the inner core, and hence, on the power required to engage the dynamo mechanism). The quasi-geostrophic motion in the layer depends on the character and intensity of interaction between the liquid layer and its boundaries.

Reference

1. Kholodova, S.E., Peregudin, S.I.: Modeling and Analysis of Streams and Waves in Liquid and Granular Mediums. St. Peterb. Gos. Univ., St. Petersburg (2009)

Parallel Implementation
of Nonparametric Clustering Algorithm
HCA-MS on GPU Using CUDA

S. A. Rylov[✉]

Institute of Computational Technologies SB RAS, Novosibirsk, Russia
RylovS@mail.ru

Abstract. The present work explores nonparametric clustering algorithm HCA-MS. The combination of grid-based approach and Mean shift procedure allows the algorithm to discover arbitrary shaped clusters and to process large datasets, such as images. Parallel implementation of the HCA-MS algorithm on NVIDIA GPU using CUDA platform is presented. Provided experimental results on model data and multispectral images confirm the efficiency of the considered algorithm and its parallel implementation. The computation speedup on images was shown to be about 20x compared to 4 core CPU.

Keywords: Clustering · Nonparametric · Grid-based · Mean shift ·
Image segmentation · GPGPU · CUDA · Parallel computing

1 Introduction

Clustering is the task of grouping a set of objects in such way that objects in the same group are more similar to each other than to those in other groups (clusters). Clustering of large datasets is urgent in many applied problems of data analysis. For example, it is one of the most common approaches to satellite image segmentation [1]. Generally, a priori information about the probabilistic characteristics of classes, as well as training samples is often absent. Widely used clustering algorithms (k-means, ISODATA, EM) are based on the assumption of Gaussian distribution models and do not always provide required segmentation quality (especially for high spatial resolution satellite images) [2,3].

In such circumstances, nonparametric clustering methods are more attractive because of their ability to discover arbitrary shaped clusters [3]. However, high computational complexity strongly limits its application to large datasets, such as images. One of the best-known nonparametric mode-seeking algorithms that are capable of producing accurate results is Mean shift [4] and it has quadratic time complexity [5].

On the other hand, grid-based methods that divide feature space into a finite number of cells are also capable of discovering arbitrary shaped clusters and at

© Springer Nature Switzerland AG 2019
Y. Shokin and Z. Shaimardanov (Eds.): CITech 2018, CCIS 998, pp. 189–196, 2019.
https://doi.org/10.1007/978-3-030-12203-4_19

the same time they have high computational efficiency [6]. But their accuracy is limited by a grid structure [7].

Recently HCA-MS algorithm was proposed by the author to combine the best qualities of these two approaches [8]. It is based on the grid-based algorithm HCA with the following Mean shift procedure for clusters' borders refinement. The combination of these approaches allows obtaining higher clustering accuracy in comparison with grid-based approach and higher performance in comparison with Mean shift.

Today, most personal computers are equipped with graphics cards (GPU), the performance of which has grown dramatically in the last years. Consequently, general-purpose computing on graphics processing units (GPGPU) technologies are actively developing for solving time-consuming non graphics tasks [9].

The present work introduces a parallel implementation of HCA-MS clustering algorithm on GPU using CUDA platform. Experimental results on model data and multispectral images confirm the efficiency of the considered algorithm and its parallel implementation. The computation speedup on images was shown to be about 20x compared to 4 core CPU, where the parallel version of the algorithm for CPU was implemented with OpenMP.

2   Nonparametric Clustering Algorithm HCA-MS

This section briefly describes the nonparametric clustering algorithm HCA-MS [8]. This novel algorithm utilizes grid-based HCA clustering algorithm with Mean shift procedure, which is applied to the elements of border cells resulting in the refinement of the clusters' borders.

The first stage of the HCA-MS is to execute grid-based HCA algorithm, which was presented in [10] and briefly described below.

Let the set of objects X consist of d-dimensional vectors lying in the feature space R^d: $X = \{x_i = (x_i^1, \ldots, x_i^d) \in R^d, i = 1, \ldots, N\}$, and bounded by a hyper-rectangle $\Omega = [l^1, r^1] \times \cdots \times [l^d, r^d]$: $l^j = \min_{x_i \in X} x_i^j$, $r^j = \max_{x_i \in X} x_i^j$. Grid structure is formed by dividing Ω with hyperplanes $x^j = (r^j - l^j) \cdot i/m + l^j$, $i = 0, \ldots, m$ where m is the number of partitions in each dimension. The set of cells adjacent to B will be denoted by A_B. The *density* D_B of the cell B is defined as the number of elements from the set X belonging to the cell B.

The nonempty cell B_i is directly connected to the nonempty cell B_j ($B_i \to B_j$) if B_j is the cell with the maximum number that satisfies the conditions $B_j = \arg \max_{B_k \in A_{B_i}} D_{B_k}$ and $D_{B_j} \geqslant D_{B_i}$. The nonempty adjacent cells B_i and B_j are directly connected ($B_i \leftrightarrow B_j$) if $B_i \to B_j$ or $B_j \to B_i$. The nonempty cells B_i and B_j are connected ($B_i \approx B_j$) if there exist $k_1, \ldots, k_l$ such that $k_1 = i$, $k_l = j$ and for all $p = 1, \ldots, l - 1$ we have $B_{k_p} \leftrightarrow B_{k_{p+1}}$. The introduced connectedness relation leads to the natural partition of nonempty cells into the *connectedness components* $\{G_1, \ldots, G_S\}$. The connected component is defined as the maximum set of pairwise connected cells. *Representative cell* $Y(G)$ of the component G is defined as a cell with the maximum number that satisfies the condition $Y(G) = \arg\max_{B \in G} D_B$.

The determined connectedness components correspond to single-mode clusters, and their representative cells correspond to the density modes of these clusters.

Next, to construct an hierarchy between components we define the distance h_{ij} between adjacent components G_i and G_j as

$$h_{ij} = \min_{P_{ij} \in \mathfrak{R}_{ij}} \left[1 - \min_{B_{k_t} \in P_{ij}} D_{B_{k_t}} \Big/ \min \left(D_{Y(G_i)}, D_{Y(G_j)} \right) \right],$$

where $\mathfrak{R}_{ij} = \{P_{ij}\}$ is a set of all possible paths between representative cells $Y(G_i)$ and $Y(G_j)$, $P_{ij} = \langle Y(G_i) = B_{k_1}, \ldots, B_{k_t}, B_{k_{t+1}}, \ldots, B_{k_t} = Y(G_j) \rangle$ such that for all $t = 1, \ldots, l - 1$: (1) $B_{k_t} \in G_i \cup G_j$; (2) $B_{k_t}, B_{k_{t+1}}$ are adjacent cells.

After forming a matrix of distances between adjacent components $\{h_{ij}\}$, the SLINK (nearest neighbor) algorithm for dendrogram construction is applied to it. The result of the algorithm is an hierarchical structure built on the set of connected components.

The HCA algorithm at low computational costs allows distinguishing clusters of complex shape and obtaining hierarchical clustering structure. Moreover, unlike the other well-known hierarchical algorithms [11], it allows separating clusters that are intersecting in the feature space. However, accuracy of cluster's separation highly depends on the grid structure, which can lead to mistakes, particularly if the grid parameter m is chosen unsuccessfully.

At the second stage of HCA-MS algorithm, data elements are grouped due to the cells they belong to, for the following quick access to the list of elements of an arbitrary cell.

At the third stage, non-empty cells located at the cluster's borders are considered. To each element of such cell Mean shift procedure is applied [4], which iteratively converges to the local density maximum: $x_{k+1} = m(x_k)$, where

$$m(x) = \frac{\sum\limits_{i=1}^{N} x_i \cdot K_{\mathrm{Ep}}(x - x_i)}{\sum\limits_{i=1}^{N} K_{\mathrm{Ep}}(x - x_i)}.$$

where the finite Epanechnikov kernel function is used:

$$K_{\mathrm{Ep}} \left(\frac{x - x_i}{h} \right) = \left(1 - \frac{||x - x_i||^2}{h^2} \right) \cdot I \left(||x - x_i|| \leqslant h^2 \right),$$

where $I(x)$—an indicator function.

Smoothing parameter h is set equal to the width of the cell in the grid structure.

The shift process stops if the considered element moves to the other non-empty cell. In case the new cell belongs to another cluster the element is moved to this cluster. The maximum number of Mean shift iterations is limited by the parameter, which we will set to 3. Experimental studies on model data have shown that in most cases this parameter value is sufficient [8]. In general, it is possible to use more sophisticated stopping criteria, but since this is not the subject of this study, we will not consider them.

3   Parallel Implementation of the Clustering Algorithm HCA-MS with CUDA

CUDA is a powerful parallel computing platform and API model that allows using NVIDIA GPUs for general purpose processing. Modern GPUs contain thousands of cores, grouped into blocks under the control of multiprocessors. The cores of one block perform the same set of instructions, but on different data elements. Each core contains small number of registers and has quick access to a limited amount of shared memory within its block (managed cache). In addition, all the threads (executed on separate cores) can access large amount of global memory, but random access time to it is slow and takes hundreds of cycles. Synchronization of threads during execution is possible only within a block.

Parallel implementation of the grid hierarchical HCA algorithm on GPU using CUDA is described in detail in [12]. By itself, this algorithm has very fast performance: four-bands images of the size up to 100 megapixels are processed on CPU within 1 s and on GPU within 0.1 s. Considering HCA-MS algorithm, its first stage (HCA) computing time is insignificant.

At the second stage, data elements are sorted by the cells and an array of the first elements indices for each cell is formed. This stage is also not computationally time consuming, and its execution is performed on CPU.

The use of a weighting table (the number of elements with the same feature values) can significantly reduce computational cost of processing color images with common 256 quantization levels. Yet, when processing multispectral satellite images containing more than three spectral bands with radiometric resolution of 10–14 bits, the efficiency of this approach is diminished. Therefore, in this particular study we will not use a weighting table.

Finally, the third stage is the one that is time consuming. Mean shift processing of the border cells can be done independently from each other, therefore, parallelization is applicable. On CPU different threads just process different cells. But, on GPU each cell is processed by its block, where block's threads process the elements of the cell in parallel. Let's consider the processing scheme of a non-empty cell by a block of threads.

First, the connectedness components of all neighboring cells are read into the shared memory by the threads, as well as the indices of the first and the last elements of these cells. If among the neighboring cells no cell belongs to the different component, which means it is not a border cell, then the block finishes its work.

Otherwise, the elements of the cell are processed in parallel by the threads. To execute Mean shift step each thread goes through all the elements from the adjacent cells to search those in the radius h. Due to the fact that it is impossible to guarantee that the number of considered elements will not exceed the limited size of shared memory, data access is made through the global memory.

Yet, to optimize data access, managed cache was successfully utilized. The elements of each cell are read into the allocated array in shared memory by fragments of fixed size that is equal to the block size (the number of threads in the block). After the threads finish processing this part of data, the next data fragment is uploaded. All block threads upload new data fragment synchronously, even if some threads are not involved in the cell elements processing. This optimization additionally reduced computation time by 20%.

4   Experimental Results

This section presents the results of experimental studies of HCA-MS algorithm on model data and images. It was shown that HCA algorithm is capable of distinguishing clusters of complex shape and produces better results than well-known clustering algorithms like K-means, EM, DBSCAN, OPTICS, DeLiClu, SLINK and Mean shift on complex data [12]. The considered issue is the efficiency of the proposed implementation to correct border mistakes caused by the grid effect. The clustering accuracy of HCA-MS algorithm is compared with the initial grid-based algorithm HCA and density-based algorithm Mean shift. The computation time of the proposed CUDA implementation of HCA-MS algorithm on GPU (GeForce GXT 770, 1536 cores) compared with execution on one and four cores of the CPU (Intel Core i5, 3.5 GHz) is presented. The parallel version of the algorithm for CPU was implemented with OpenMP.

Experiment 1. Two-dimensional synthetic dataset containing 3 classes [13] was clustered. One cluster described by a normal distribution is surrounded by two clusters in the form of rings (Fig. 1c). Mean shift clustering algorithm cannot extract multimode clusters in the form of rings in principle. In its turn, HCA algorithm successfully separates all three clusters, but makes some mistakes when the grid parameter is selected unsuccessfully: at $m = 30$ clustering accuracy is 98.21% (Fig. 1a); at $m = 42$ mistakes are made in 3 points (Fig. 1b); 100% accuracy is obtained only at $m = 46$ (Fig. 1c). However, HCA-MS algorithm is able to correct mistakes caused by the grid effect. As a result, 100% accuracy is obtained in all three cases: at $m = 30$, $m = 42$ and $m = 46$ (Fig. 1c).

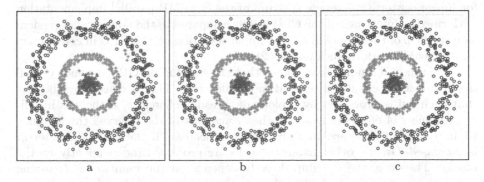

Fig. 1. The results of clustering the model dataset obtained by HCA algorithm at $m = 30$ (a), $m = 42$ (b), $m = 46$ (c); and the result of HCA-MS with the same parameter's values (c).

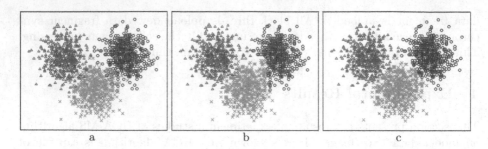

a   b   c

Fig. 2. Clustering results of the model dataset (a) by HCA algorithm (b) and by HCA-MS algorithm (c) at $m = 20$.

Experiment 2. The synthetic dataset containing three strongly intersecting classes with normal distribution [13] is shown in Fig. 2a. Clustering accuracy of this model by HCA algorithm at $m = 20$ is 94.93% (Fig. 2b). At this case, the accuracy can be increased by using finer grid: for example, at $m = 40$, the accuracy is 96%. However, very fine grid can be unacceptable when extracting clusters with complex structure. HCA-MS algorithm demonstrates 96.4% accuracy at $m = 20$ (Fig. 2c). While Mean-shift algorithm at most achieves 96.33% accuracy at $h = 27$. At the same time, other density-based algorithms (DBSCAN, OPTICS, DeLiClu) fail to adequately separate strongly intersecting classes.

Experiment 3. To assess the speedup effect of the GPU implementation, test data-sets of different size were generated. The data-sets follow uniform distribution in the limited three-dimensional feature space. HCA-MS algorithm processing time on the test data depending on the number of elements (from 500'000 to 10'000'000) performed on the GPU and one and four cores of the CPU is presented in Fig. 3a. The obtained speedup of the GPU execution in comparison with four cores of the CPU is shown in Fig. 3b. The results are presented for the grid parameter values $m = 20$ and $m = 32$, because these values are usually used for image clustering. Average speedup of the OpenMP parallel implementation performed on four cores of the CPU is 3.8x compared to the non-parallel version. The speedup of the proposed CUDA implementation on the GPU in comparison with OpenMP performance reaches 28x.

Experimental studies have shown that the speedup on GPU depends on cell density: denser cells are processed more efficiently. This is caused by the fact that each cell is processed by a block of threads. Therefore, if there is insufficient number of elements in a cell, then some part of the reserved threads may be unused. Thus, the speedup on GPU can decrease if the grid parameter m is increased. On the other hand, the cells are processed independently by the blocks. Therefore, the speedup directly depends on the number of streaming multiprocessors (SM). And while the number of cores per SM is almost constant (192 in Kepler and 128 in Maxwell and Pascal architectures), the number of SMs is determined by the total number of GPU cores, which provides a direct dependence of the algorithm CUDA performance on the number of GPU cores.

Experiment 4. The table below shows HCA-MS algorithm time performance on the GPU and four cores of the CPU on the images of different size at $m = 20$. In total 40 color photos and multispectral satellite images (WorldView-2 and Landsat-8) of different size [14] were processed. The full table is avable here [15]. The results showed that the average speedup of the OpenMP parallel implementation performed on four cores of the CPU is 3.7x compared to the non-parallel version and it does not fluctuate considerably. The average speedup of the GPU execution in comparison with four cores of the CPU is 22x at $m = 20$ and 19x at $m = 32$ (for the images containing more than 1 million pixels). Time performance of the algorithm at $m = 32$ is about 4x faster than at $m = 20$. However, it should be noticed that processing time strongly depends on the data itself, and for different images of the same size it can vary greatly.

Thus, the proposed parallel implementation on the GPU can process large multispectral images just in few minutes (Table 1).

Table 1. HCA-MS algorithm time performance on the images (in seconds).

Number of bands	3	3	3	3	4	3	3	4	3	3	4
Image size (megapixels)	0.3	1.2	4.2	4.2	4.2	5	9	12.5	13.8	25	25
CPU, 4 cores	6.6	93	302	6872	701	1465	1948	2055	18186	16623	11013
GPU	0.5	5.2	10.6	357	33	49	96	115	679	587	510
Speedup	14.4	18.0	28.4	19.2	21.3	29.7	20.3	17.8	26.8	28.3	21.6

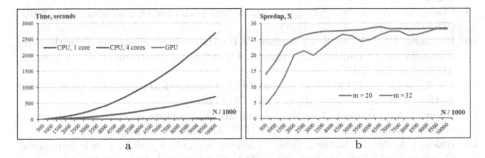

a   b

Fig. 3. HCA-MS algorithm processing time on the test data depending on the number of elements N performed on the GPU and one and four cores of the CPU (a); the speedup of the GPU performance in comparison with 4 cores of the CPU (b).

5   Conclusion

Parallel implementation of the nonparametric HCA-MS clustering algorithm on GPU using CUDA platform is presented. Experimental results on model datasets showed the ability of the algorithm to correct mistakes caused by the grid effect, reaching clustering accuracy level of the well-known Mean shift algorithm. The experiments showed, that the proposed parallel implementation on GPU allows processing multispectral images 20 times faster than on CPU (4 cores). Thereby, large multispectral images can be clustered just in few minutes.

References

1. Xie, Y., Sha, Z., Yu, M.: Remote sensing imagery in vegetation mapping: a review. J. Plant Ecol. **1**(1), 9–23 (2008)
2. Zadkarami, M.R., Rowhani, M.: Application of skew-normal in classification of satellite image. J. Data Sci. **8**, 597–606 (2010)
3. Sarmah, S., Bhattacharyya, D.K.: A grid-density based technique for finding clusters in satellite image. Pattern Recogn. Lett. **33**(5), 589–604 (2012)
4. Cheng, Y.: Mean shift, mode seeking, and clustering. IEEE Trans. Pattern Anal. Mach. Intell. **17**(8), 790–799 (1995)
5. Freedman, D., Kisilev, P.: Fast mean shift by compact density representation. In: IEEE Conference on Computer Vision and Pattern Recognition (CVPR), pp. 1818–1825. IEEE (2009)
6. Ilango, M.R., Mohan, V.: A survey of grid based clustering algorithms. Int. J. Eng. Sci. Technol. **2**(8), 3441–3446 (2010)
7. Krstinic, D., Skelin, A.K., Slapnicar, I.: Fast two-step histogram-based image segmentation. IET Image Process. **5**(1), 63–72 (2011)
8. Rylov, S.A.: Nonparametric clustering algorithm for image segmentation combining grid-based approach and mean-shift procedure. In: CEUR Workshop Proceedings, vol. 2033, pp. 150–155 (2017)
9. Choquette, J., Giroux, O., Foley, D.: Volta: performance and programmability. IEEE Micro **38**(2), 42–52 (2018)
10. Pestunov, I.A., Rylov, S.A., Berikov, V.B.: Hierarchical clustering algorithms for segmentation of multispectral images. Optoelectron. Instrument. Data Process. **51**(4), 329–338 (2015)
11. Lu, Y., Wan, Y.: PHA: a fast potential-based hierarchical agglomerative clustering method. Pattern Recogn. **46**(5), 1227–1239 (2013)
12. Rylov, S.A., Pestunov, I.A.: Fast hierarchical clustering of multispectral images and its implementation on NVIDIA GPU. JPCS **1096**, 012039 (2018)
13. Rylov, S.A.: Model datasets. [Electronic resource]. https://drive.google.com/open?id=0ByK9GtU5ExExRnZwdFNmRHRWdFk. Accessed 20 Apr 2018
14. Image datasets for clustering. [Electronic resource]. https://drive.google.com/open?id=0ByK9GtU5ExExWXpGRjU5WVFHcDg. Accessed 20 Apr 2018
15. Table: HCA-MS algorithm time performance on the images. [Electronic resource]. https://drive.google.com/file/d/1xA89kC3tixwEUMLaX_pfEOTJ-s-Uh0-8/view?usp=sharing Accessed 14 Oct 2018

Numerical Algorithm for Solving the Inverse Problem for the Helmholtz Equation

M. A. Shishlenin[1,2,3(✉)], S. E. Kasenov[4], and Zh. A. Askerbekova[4]

[1] Institute of Computational Mathematics and Mathematical Geophysics SB RAS, Novosibirsk, Russia
mshishlenin@ngs.ru
[2] Novosibirsk State University, Novosibirsk, Russia
[3] Sobolev Institute of Mathematics, Novosibirsk, Russia
[4] Al-Farabi Kazakh National University, Almaty, Kazakhstan

Abstract. In this paper we consider acoustic equation. The equation by separation of variables is reduced to a boundary value problem for the Helmholtz equation. We consider problem for the Helmholtz equation. We reduce the solution of the operator equation to the problem of minimizing the functional. And we build numerical algorithm for solving the inverse problem. At the end of the article is given the numerical calculations of this problem.

Keywords: Continuation problem · Regularization problem ·
Comparative analysis · Numerical methods · Landweber's method

1 Introduction

For mathematical modelling of physical processes and the phenomena occurring in nature, it is necessary to face ill-posed problems, including with the Cauchy problem for the Helmholtz equation. The Helmholtz equation is used in many physical processes associated with the propagation of waves and has numerous applications. If the law of oscillations of the physical medium harmonically depends on time, then the wave equation can be transformed to the Helmholtz equation. In particular, the Cauchy problem for the Helmholtz equation describes the propagation of electromagnetic or acoustic waves. The aim of the paper is that an effective numerical solution for investigating inverse elliptic-type problems by the Landweber method. A significant theoretical and applied contribution to this topic has been accumulated in monographs by A.N. Tikhonova, M.M. Lavrentyeva, V.K. Ivanova, A.V. Goncharsky. The Cauchy problem for elliptic equations is of fundamental importance in all inverse problems. An important application of the Helmholtz equation is the acoustic wave problem, which is considered in the works of DeLillo, Isakov, Valdivia, Wang (2003) L. Marin, L. Elliott, P. J. Heggs, D. B. Ingham, D. Lesnic and H. Wen. The Landweber

© Springer Nature Switzerland AG 2019
Y. Shokin and Z. Shaimardanov (Eds.): CITech 2018, CCIS 998, pp. 197–207, 2019.
https://doi.org/10.1007/978-3-030-12203-4_20

method is effective and makes it possible to substantially simplify the investigation of inverse problems [1,2].

There are a lot of application of the Cauchy problems for PDE [3–5]. In the work [6] authors introduced a concept of very weak solution to a Cauchy problem for elliptic equations. The Cauchy problem is regularized by a well-posed non-local boundary value problem the solution of which is also understood in a very weak sense. A stable finite difference scheme is suggested for solving the non-local boundary value problem and then applied to stabilizing the Cauchy problem. Numerical examples are presented for showing the efficiency of the method.

In the work [7] it was investigated the ill-posedness of the Cauchy problem for the wave equation. The conditional-stability estimate was proved.

In [8] it was investigated the continuation problem for the elliptic equation. The continuation problem is formulated in operator form $Aq = f$. The singular values of the operator A are presented and analyzed for the continuation problem for the Helmholtz equation. Results of numerical experiments are presented.

2   Formulation of the Problem

Consider the acoustics equation [10] in domain $Q = \Omega \times (0, +\infty)$, where $\Omega = (0, 1) \times (0, 1)$:

$$c^{-2}(x, y)U_{tt} = \Delta U - \nabla \ln(\rho(x, y))\nabla U \qquad (x, y, t) \in Q \qquad (1)$$

Suppose that a harmonic oscillation regime was established in Ω:

$$U(x, y, t) = u(x, y)e^{i\omega t}, \qquad (x, y, t) \in Q \qquad (2)$$

Putting (2) into (1) we obtain Helmholtz equation:

$$-\omega^2 c^{-2}u = \Delta u - \nabla \ln(\rho(x, y))\nabla u, \qquad (x, y) \in \Omega$$

We consider the initial-boundary value problem:

$$
\begin{align}
-\omega^2 c^{-2}u &= \Delta u - \nabla \ln(\rho(x, y))\nabla u, & (x, y) &\in \Omega, & (3) \\
u(0, y) &= h_1(y), & y &\in [0, 1], & (4) \\
u(x, 0) &= h_2(x), & x &\in [0, 1], & (5) \\
u_x(0, y) &= f_1(y), & y &\in [0, 1], & (6) \\
u_y(x, 0) &= f_2(x), & x &\in [0, 1]. & (7)
\end{align}
$$

Problem (3)–(7) appears ill-posed. For a numerical solution of the problem, we first reduce it to the inverse problem $Aq = f$ with respect to some direct (well-posed) problem. Further, we reduce the solution of the operator equation $Aq = f$ to the problem of minimizing the objective functional $J(q) = \langle Aq - f, Aq - f \rangle$. After calculating the gradient $J'q$ of the objective functional, we apply the method of Landweber to minimize it [11,12].

3   The Conditional Stability Theorem

Let us consider the initial-boundary value problem:

$$\Delta u = 0, \qquad\qquad\qquad\qquad\qquad\qquad (x,y) \in \Omega, \qquad (8)$$
$$u(0,y) = f_1(y), \quad u_x(0,y) = h_1(y), \qquad\qquad y \in [0,1], \qquad (9)$$
$$u(x,0) = f_2(x), \quad u_y(x,0) = h_2(x), \qquad\qquad x \in [0,1], \qquad (10)$$

Let us divide the problem into two parts:

Problem 1	Problem 2
$\Delta u = 0,$	$\Delta u = 0,$
$u(0,y) = f_1(y),$	$u(0,y) = 0,$
$u(x,0) = 0,$	$u(x,0) = f_2(x),$
$u_x(0,y) = h_1(y),$	$u_x(0,y) = 0,$
$u_y(x,0) = 0.$	$u_y(x,0) = h_2(x).$

Problem 1, we continue the field along the axis x, then at $y = 1$ we can admit the boundary at zero. And also, problem 2, we continue the field along the axis y, then at $x = 1$ we can admit the boundary at zero. Suppose $h_2(x) = 0, h_1(y) = 0$.

Problem 1

$$\Delta u = 0, \qquad\quad (x,y) \in \Omega, \quad (11)$$
$$u(0,y) = f_1(y), \quad y \in [0,1], \quad (12)$$
$$u(x,0) = 0, \qquad\quad x \in [0,1], \quad (13)$$
$$u_x(0,y) = 0, \qquad y \in [0,1], \quad (14)$$
$$u(x,1) = 0, \qquad\quad x \in [0,1]. \quad (15)$$

Problem 2

$$\Delta u = 0, \qquad\quad (x,y) \in \Omega, \quad (16)$$
$$u(0,y) = 0, \qquad\quad y \in [0,1], \quad (17)$$
$$u(x,0) = f_2(x), \quad x \in [0,1], \quad (18)$$
$$u(1,y) = 0, \qquad\quad y \in [0,1], \quad (19)$$
$$u_y(x,0) = 0, \qquad x \in [0,1]. \quad (20)$$

Theorem 1 (of the conditional stability). *Let us suppose that for $f_1 \in L_2(0,1)$ and there is a solution $u \in L_2(\Omega)$ of the problem (11)–(15). Then the following estimate of conditional stability is right*

$$\int_0^1 u^2(x,y)dy \le \left(\int_0^1 f_1^2(y)dy \right)^{1-x} \left(\int_0^1 u^2(1,y)dy \right)^x. \qquad (21)$$

Theorem 2 (of the conditional stability). *Let us suppose that for $f_2 \in L_2(0,1)$ and there is a solution $u \in L_2(\Omega)$ of the problem (16)–(20). Then the following estimate of conditional stability is right*

$$\int_0^1 u^2(x,y)dx \le \left(\int_0^1 f_2^2(x)dx \right)^{1-y} \left(\int_0^1 u^2(x,1)dx \right)^y. \qquad (22)$$

More details proof such estimates are shown in works [13,14].

4 Reduction of the Initial Problem to the Inverse Problem

Let us show that the solution of the problem (3)–(7) is possible to reduce to the solution of the inverse problem with respect to some direct (well-posed) problem [?], [?].

As a direct problem, we consider the following one

$$
\begin{aligned}
-\omega^2 c^{-2} u &= \Delta u - \nabla \ln(\rho(x,y)) \nabla u, & (x,y) &\in \Omega, & (23) \\
u(0,y) &= h_1(y), & y &\in [0,1], & (24) \\
u(x,0) &= h_2(x), & x &\in [0,1], & (25) \\
u(1,y) &= q_1(y), & y &\in [0,1], & (26) \\
u(x,1) &= q_2(x), & x &\in [0,1]. & (27)
\end{aligned}
$$

The inverse problem to problem (23)–(27) consist in defining the function $q_1(x), q_2(y)$ by the additional information on the solution of direct problem.

$$
\begin{aligned}
u_x(0,y) &= f_1(y), & y &\in [0,1], & (28) \\
u_y(x,0) &= f_2(x), & x &\in [0,1]. & (29)
\end{aligned}
$$

We introduce the operator

$$
A\colon (q_1, q_2) \mapsto (u_x(0,y), u_y(x,0)). \tag{30}
$$

Then the inverse problem can be written in operator form

$$
Aq = f.
$$

We introduce the objective functional

$$
J(q_1, q_2) = \int_0^1 \left[u_x(0,y; q_1, q_2) - f_1(y) \right]^2 dy + \int_0^1 \left[u_y(x,0; q_1, q_2) - f_2(x) \right]^2 dx. \tag{31}
$$

We shall minimize the quadratic functional (31) by Landweber's method. Let the approximation be known q^n. The subsequent approximation is determined from:

$$
q^{n+1} = q^n - \alpha J'(q^n) \tag{32}
$$

here $\alpha \in (0, \|A\|^{-2})$ [10].

Let us note that the convergence of the Lanweber iteration can be sufficiently increase if we apply the apriori information about the solution [9].

Algorithm for Solving the Inverse Problem

1. We choose the initial approximation $q^0 = (q_1^0, q_2^0)$;
2. Let us assume that q_n is known, then we solve the direct problem numerically

$$u_{xx} + u_{yy} - \left(\frac{\rho_x}{\rho}u_x + \frac{\rho_y}{\rho}u_y\right) + \left(\frac{\omega}{c}\right)^2 u = 0, \qquad (x,y) \in \Omega,$$

$$u(0,y) = h_1(y), \quad u(1,y) = q_1^n(y), \qquad\qquad y \in [0,1],$$
$$u(x,0) = h_2(x), \quad u(x,1) = q_2^n(x), \qquad\qquad x \in [0,1].$$

3. We calculate the value of the functional

$$J(q_{n+1}) = \int\limits_0^1 \left[u_x(0,y;q_1^{n+1}) - f_1(y)\right]^2 dy + \int\limits_0^1 \left[u_y(x,0;q_2^{n+1}) - f_2(x)\right]^2 dx;$$

4. If the value of the functional is not sufficiently small, then go to next step;
5. We solve the conjugate problem

$$\psi_{xx} + \psi_{yy} + \left(\frac{\rho_x}{\rho}\psi\right)_x + \left(\frac{\rho_y}{\rho}\psi\right)_y + \left(\frac{\omega}{c}\right)^2 \psi = 0, \qquad (x,y) \in \Omega,$$

$$\psi(0,y) = 2\big(u_x(0,y;q_1) - f_1(y)\big), \psi(1,y) = 0, \qquad y \in [0,1],$$
$$\psi(x,0) = 2\big(u_y(x,0;q_2) - f_2(x)\big), \psi(x,1) = 0, \qquad x \in [0,1].$$

6. Calculate the gradient of the functional $J'(q^n) = \big(-\psi_x(1,y), -\psi_y(x,1)\big)$;
7. Calculate the following approximation $q^{n+1} = q^n - \alpha J'(q^n)$, then turn to step 2.

5   Numerical Solution of the Inverse Problem

First we consider the initial problem in a discrete statement. We carry out a numerical study of the stability of the problem in a discrete statement [?].

Discretization of the Original Problem
The corresponding difference problem for the original problem (3)–(7) has the following

$$\frac{u_{i+1,j} - 2u_{i,j} + u_{i-1,j}}{h^2} + \frac{u_{i,j+1} - 2u_{i,j} + u_{i,j-1}}{h^2}$$

$$- \frac{\rho_{i+1,j} - \rho_{i-1,j}}{2h\rho_{i,j}} \cdot \frac{u_{i+1,j} - u_{i-1,j}}{2h}$$

$$- \frac{\rho_{i,j+1} - \rho_{i,j-1}}{2h\rho_{i,j}} \cdot \frac{u_{i,j+1} - u_{i,j-1}}{2h} + \left(\frac{\omega}{c}\right)^2 u_{i,j} = 0, \qquad i,j = \overline{1, N-1},$$

$$u_{0,j} = h_1^j, \qquad\qquad\qquad\qquad\qquad j = \overline{0, N},$$
$$u_{i,0} = h_2^i, \qquad\qquad\qquad\qquad\qquad i = \overline{0, N},$$
$$u_{1,j} = h_1^j + h \cdot f_1^j, \qquad\qquad\qquad\qquad j = \overline{0, N},$$
$$u_{i,1} = h_2^i + h \cdot f_2^i, \qquad\qquad\qquad\qquad i = \overline{0, N}.$$

For convenience, we introduce the new denotations $a_{i,j} = 1 + \dfrac{\rho_{i+1,j} - \rho_{i-1,j}}{4\rho_{i,j}}$,

$b_{i,j} = 1 + \dfrac{\rho_{i,j+1} - \rho_{i,j-1}}{4\rho_{i,j}}$, $\quad c = -4 + \left(\dfrac{\omega \cdot h}{c}\right)^2$, $\quad d_{i,j} = 1 - \dfrac{\rho_{i+1,j} - \rho_{i-1,j}}{4\rho_{i,j}}$,

$e_{i,j} = 1 - \dfrac{\rho_{i,j+1} - \rho_{i,j-1}}{4\rho_{i,j}}$.

$$a_{i,j}u_{i-1,j} + b_{i,j}u_{i,j-1} + cu_{i,j} + d_{i,j}u_{i,j+1} + e_{i,j}u_{i+1,j} = 0, \quad i,j = \overline{1, N-1},$$
(33)

$$u_{0,j} = h_1^j, \qquad\qquad\qquad\qquad\qquad\qquad\qquad\qquad j = \overline{0, N},$$
(34)

$$u_{i,0} = h_2^i, \qquad\qquad\qquad\qquad\qquad\qquad\qquad\qquad i = \overline{0, N},$$
(35)

$$u_{1,j} = h_1^j + h \cdot f_1^j,$$
(36)

$$u_{i,1} = h_2^i + h \cdot f_2^i, \qquad\qquad\qquad\qquad\qquad\qquad\qquad i = \overline{0, N}.$$
(37)

Let us construct a system of difference equations [15, p. 379]

$$A \cdot X = B.$$
(38)

Here A—of matrix $(N+1)^2$ size, X—unknown vector of the form

$$X = (u_{0,0}, u_{0,1}, u_{0,2} \ldots u_{0,N}, u_{1,0}, u_{1,1}, u_{1,2} \ldots u_{1,N}, \ldots u_{N,0}, u_{N,1}, u_{N,2}, \ldots u_{N,N}),$$

B—data vector (boundary and additional conditions).

Analysis of the Stability of the Matrix of the Initial Problem

Description of the numerical experiment $c = 1, \quad \omega = 0.5$

$$h_1(y) = \frac{1 - \cos(8\pi y)}{4}, \quad h_2(x) = \frac{1 - \cos(8\pi x)}{4},$$

$$q_1(y) = \frac{1 - \cos(8\pi y)}{4}, \quad q_2(x) = \frac{1 - \cos(8\pi x)}{4},$$

$$\rho(x,y) = e^{-\frac{(x-0.5)^2 + (y-0.5)^2}{2b^2}}, \quad b = 0.1.$$

Table 1 presents the results of a singular decomposition of the matrix of the initial problem A and a direct problem A_T for the values $N = 50$

Table 1. Singular decomposition of matrices with size $(N+1)^2$

Matrices	$\sigma_{max}(A)$	$\sigma_{min}(A)$	$\mu(A)$
A_T	743.404	0.015	47056.2
A	743.404	$9.07 \cdot 10^{-19}$	$8.19 \cdot 10^{20}$

The matrix of the original problem has a poor conditionality [16] (Table 2).

Numerical Results of the Inverse Problem by the Landweber Method
In this section, to solve the two-dimensional direct problem for the Helmholtz equation, the finite element method is used. Triangulation with the number of triangles—N_t; vertices—N_v; and the number of points at the border—N. The problem is solved using the computational package FreeFEM++ (Figs. 1, 2, 3, 4 and 5).

Description of the numerical experiment $c = 1, \quad \omega = 0.5$

$$h_1(y) = \frac{1 - \cos(8\pi y)}{4}, \quad h_2(x) = \frac{1 - \cos(8\pi x)}{4},$$

$$q_1(y) = \frac{1 - \cos(8\pi y)}{4}, \quad q_2(x) = \frac{1 - \cos(8\pi x)}{4},$$

$$\rho(x, y) = e^{-\frac{(x-0.5)^2 + (y-0.5)^2}{2b^2}}, \quad b = 0.1.$$

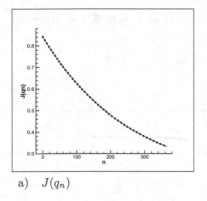

a) $J(q_n)$

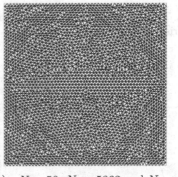

b) $N = 50, N_t = 5862$ and $N_v = 3032$

Fig. 1. (a) The value of the functional by iteration, (b) Ω area grid with N number of points on the border

Table 2. Solution results by the Landweber iteration method without noise

Number of iterations, n	$J(q)$	$\|u_T - \tilde{u}\|$
10	0.8158	0.1491
100	0.6254	0.1013
300	0.3788	0.0553
365	0.3323	0.0538

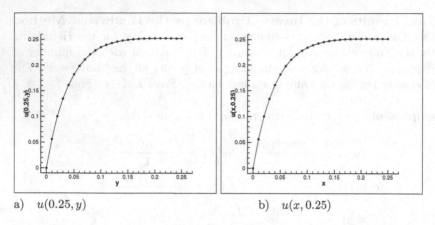

a) $u(0.25, y)$ b) $u(x, 0.25)$

Denotation: (symbol ■) — Landweber solution, (symbol ●) — exact solution

Fig. 2. The figure (a) comparison of boundaries $u(x, y)$ at $x = 0.25$, the figure (b) comparison of boundaries $u(x, y)$ at $y = 0.25$

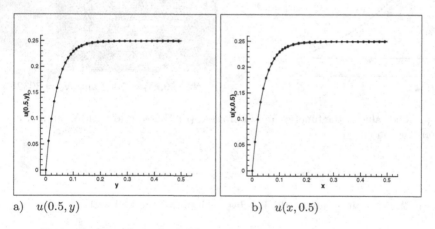

a) $u(0.5, y)$ b) $u(x, 0.5)$

Denotation: (symbol ■) — Landweber solution, (symbol ●) — exact solution

Fig. 3. The figure (a) comparison of boundaries $u(x, y)$ at $x = 0.5$, the figure (b) comparison of boundaries $u(x, y)$ at $y = 0.5$

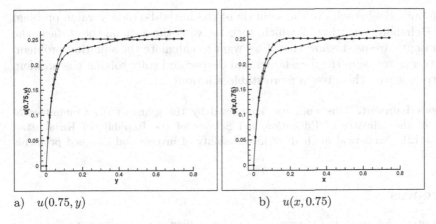

a) $u(0.75, y)$ b) $u(x, 0.75)$

Denotation: (symbol ■) — Landweber solution, (symbol •) — exact solution

Fig. 4. The figure (a) comparison of boundaries $u(x, y)$ at $x = 0.75$, the figure (b) comparison of boundaries $u(x, y)$ at $y = 0.75$

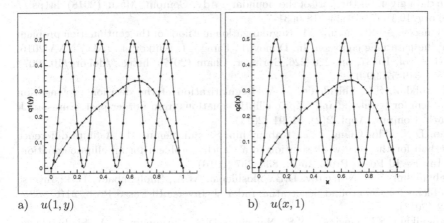

a) $u(1, y)$ b) $u(x, 1)$

Denotation: (symbol ■) — Landweber solution, (symbol •) — exact solution

Fig. 5. The figure (a) comparison of boundaries $u(x, y)$ at $x = 1$, the figure (b) comparison of boundaries $u(x, y)$ at $y = 1$

6   Conclusion

The paper is devoted to the investigation of an ill-posed problem by initial-boundary value problems for the Helmholtz equation, the construction of numerical optimization methods for solving problems, the construction of corresponding algorithms and the computational experiments of this problem.

The numerical results of the solution of the initial-boundary value problem for the Helmholtz equation, in which, together with the data on the surface, the data in depth are used, show that if we want to calculate the squaring problem, it is better to measure the data larger and deeper and start solving the problem in a large square. This gives a more stable solution.

Acknowledgement. This work was supported by the grant of the Committee of Science of the Ministry of Education and Science of the Republic of Kazakhstan (AP05134121 "Numerical methods of identifiability of inverse and ill-posed problems of natural science".

References

1. DeLillo, T., Isakov, V., Valdivia, N., Wang, L.: The detection of the source of acoustical noise in two dimensions. SIAM J. Appl. Math. **61**, 2104–2121 (2001)
2. DeLillo, T., Isakov, V., Valdivia, N., Wang, L.: The detection of surface vibrations from interior acoustical pressure. Inverse Prob. **19**, 507–524 (2003)
3. Belonosov, A., Shishlenin, M., Klyuchinskiy, D.: A comparative analysis of numerical methods of solving the continuation problem for 1D parabolic equation with the data given on the part of the boundary. Adv. Comput. Math. (2018). https://doi.org/10.1007/s10444-018-9631-7
4. Belonosov, A., Shishlenin, M.: Regularization methods of the continuation problem for the parabolic equation. In: Dimov, I., Faragó, I., Vulkov, L. (eds.) NAA 2016. LNCS, vol. 10187, pp. 220–226. Springer, Cham (2017). https://doi.org/10.1007/978-3-319-57099-0_22
5. Kabanikhin, S.I., Shishlenin, M.A.: Regularization of the decision prolongation problem for parabolic and elliptic elliptic equations from border part. Eurasian J. Math. Comput. Appl. **2**(2), 81–91 (2014)
6. Hào, D.N., Thu Giang, L.T., Kabanikhin, S., Shishlenin, M.: A finite difference method for the very weak solution to a Cauchy problem for an elliptic equation. J. Inverse Ill-Posed Prob. **26**(6), 835–857 (2018)
7. Kabanikhin, S.I., Nurseitov, D.B., Shishlenin, M.A., Sholpanbaev, B.B.: Inverse problems for the ground penetrating radar. J. Inverse Ill-Posed Prob. **21**(6), 885–892 (2013)
8. Kabanikhin, S.I., Gasimov, Y.S., Nurseitov, D.B., Shishlenin, M.A., Sholpanbaev, B.B., Kasenov, S.: Regularization of the continuation problem for elliptic equations. J. Inverse Ill-Posed Prob. **21**(6), 871–884 (2013)
9. Kabanikhin, S., Shishlenin, M.: Quasi-solution in inverse coefficient problems. J. Inverse Ill-Posed Prob. **16**(7), 705–713 (2008)
10. Kabanikhin, S.I.: Inverse and Ill-Posed Problems: Theory and Applications. De Gruyter, Germany (2012)
11. Bektemesov, M.A., Nursetov, D.B., Kasenov, S.E.: Numerical solution of the two-dimensional inverse acoustics problem. Bull. KazNPU Ser. Phys. Math. **1**(37), 47–53 (2012)
12. Reginska, T., Reginski, K.: Approximate solution of a Cauchy problem for the Helmholtz equation. Inverse Prob. **22**, 975–989 (2006)
13. Kasenov, S., Nurseitova, A., Nurseitov, D.: A conditional stability estimate of continuation problem for the Helmholtz equation. In: AIP Conference Proceedings, vol. 1759, p. 020119 (2016)

14. Kabanikhin, S.I., Shishlenin, M.A., Nurseitov, D.B., Nurseitova, A.T., Kasenov, S.E.: Comparative analysis of methods for regularizing an initial boundary value problem for the Helmholtz Equation. J. Appl. Math. (2014). Article id 786326
15. Samarsky, A.A., Gulin, A.V.: Numerical Methods. Nauka, Moscow (1989)
16. Godunov, S.K.: Lectures on Modern Aspects of Linear Algebra. Science Book, Novosibirsk (2002)

Implicit Iterative Schemes for Solving Stationary Problems of an Incompressible Fluid with a Large Margin of Stability

Ye. Yergaliyev$^{(\boxtimes)}$ and M. Madiyarov

Department of Mathematics, S. Amanzholov East Kazakhstan State University,
148, Voroshilov, 070002 Ust-Kamenogorsk, Kazakhstan
ergaliev79@mail.ru, madiyarov_mur@mail.ru

Abstract. This paper is devoted to the construction and investigation of difference schemes for equations describing the motion of a viscous incompressible fluid in natural "velocity vector - pressure" variables. Much attention is paid to the implicit difference iterative schemes developed on the basis of the idea of "weak compressibility".

Mathematical problems arising when studying the motion of a viscous incompressible fluid are of current importance both in the theoretical plan and in the study of specific models used in mechanics, physics, and other natural sciences to describe real processes. The processes associated with the flow of a viscous incompressible fluid are successfully described by the Navier-Stokes equations. These systems of equations are nonlinear, do not belong to the evolutionary Cauchy-Kovalevskaya type. The absence of a boundary condition for the pressure on the solid walls of the region under consideration, where the values for the velocity vector components and the small parameter for the higher derivatives are given also lead to technological difficulties. These circumstances certainly complicate the search for analytical solutions of such systems of equations, and with the current state of mathematics they can be solved only by computational methods.

Keywords: Iterative scheme · Numerical algorithm · Decision error · Velocity field · Isoline of the current function

1 Introduction

The Navier-Stokes equations describing a viscous incompressible fluid have attracted the attention of scientists dealing with solvability of partial differential equations and specialists in the field of numerical analysis for many decades because of their many applications.

Numerous monographs and scientific articles [1–4] have been devoted to the numerical solution of the system of differential equations of an incompressible fluid using finite difference methods and their mathematical substantiation. The

© Springer Nature Switzerland AG 2019
Y. Shokin and Z. Shaimardanov (Eds.): CITech 2018, CCIS 998, pp. 208–217, 2019.
https://doi.org/10.1007/978-3-030-12203-4_21

construction of effective numerical algorithms for solving the Navier-Stokes equations for a viscous incompressible fluid is of great interest to specialists in the field of computational fluid dynamics. When considering this system in natural variables, computational and theoretical difficulties arise primarily due to the absence of a boundary condition for the pressure on solid walls and the mathematical justification of stability, convergence, and obtaining estimates of the rate of convergence. Therefore, further development of the theory of difference methods for solving equations of incompressible fluid is a topical problem in computational mathematics.

2   Formulation of the Problem

Consider the following stationary system of the Navier-Stokes equation for an incompressible fluid in the rectangular domain $D = \{0 \leq x_\alpha \leq l_\alpha, \alpha = 1, 2, ..., N\}$:

$$(\boldsymbol{u} \cdot \nabla) \boldsymbol{u} + \operatorname{grad} p = \nu \Delta \boldsymbol{u} + \boldsymbol{f}(x), \quad x \in D,$$

$$\operatorname{div} \boldsymbol{u} = 0 \tag{1}$$

with a boundary condition

$$\boldsymbol{u}\big|_{\partial D} = 0, \tag{2}$$

where $\boldsymbol{u} = (u_1, u_2, ..., u_N)$ is the velocity vector; p is the pressure; ν is the coefficient of viscosity.

First, we study the rate of convergence of one class of iterative schemes for the numerical solution of stationary grid equations, corresponding to the difference approximation of the system (1), (2) of the form

$$L_{h,u} \boldsymbol{u} + \overline{\operatorname{grad}}_h p = \nu \Delta_h \boldsymbol{u} + \boldsymbol{f}, \tag{3}$$

$$\underline{\operatorname{div}}_h \boldsymbol{u} = 0, \quad x \in D_h \tag{4}$$

with homogeneous boundary conditions for the components of the velocity vector. Here and below we use the notation from the theory of difference schemes [1] and [5,6]. We also assume that the operators $L_{h,m}$, $m = \overline{1, N}$, corresponding to the approximation of the convective terms, are energetically "neutral", i.e.

$$(L_{h,m} u_m, u_m) = 0, \quad \forall m = \overline{1, N},$$

and the following additional condition holds:

$$\sum_{D_h} p(x) = 0, \tag{5}$$

which corresponds to the condition of uniqueness of the pressure determination. For the numerical solution of the grid stationary equations of an incompressible fluid in the *velocity vector, pressure* variables (3), (4), we consider an iterative scheme developed on the basis of the idea of "weak compressibility":

$$\frac{u_m^{n+1} - u_m^n}{\tau} + L_{h,m} u_m^{n+1} + (p^n - \tau_0 \underline{\text{div}}_h u)_{x_m} =$$

$$= \nu \Delta_h u_m^{n+1} + \tau_0 \delta \left(u_m^{n+1} - u_m^n \right)_{x_m \bar{x}_m} + f_m, \ m = 1, 2, ..., N, \tag{6}$$

$$\frac{p^{n+1} - p^n}{\tau_0} + \underline{\text{div}}_h u^{n+1} = 0 \tag{7}$$

with homogeneous boundary conditions

$$u^{n+1}\Big|_{\partial D_h} = 0, \tag{8}$$

where τ, τ_0, δ are positive iterative parameters.

The boundedness of the iteration in the cases of a linear and nonlinear problem is proved for the proposed algorithm (6)–(8), and in the case of the Stokes problem, it is revealed that the rate of convergence does not depend on the number of nodes of the finite difference grid [7,8].

The difference relation for the error of the solution in the case of the linear Stokes problem is as follows:

$$\frac{z_m^{n+1} - z_m^n}{\tau} + (\pi^n - \tau_0 \underline{\text{div}}_h z^n)_{x_m} = \nu \Delta_h z_m^{n+1} + \tau_0 \delta \left(z_m^{n+1} - z_m^n \right)_{x_m \bar{x}_m}, \tag{9}$$

$$\frac{\pi^{n+1} - \pi^n}{\tau} + \underline{\text{div}}_h z^{n+1} = 0, \tag{10}$$

where

$$z_m^n(x) = u_m^n(x) - u_m(x), x \in D_{h,m},$$

$$\pi^n = p^n - p(x), x \in D_h,$$

and the following theorem on the rate of convergence of the iterative process holds.

Theorem 1. The following estimate holds for the error of the iterative process (9), (10):

$$F^{n+1} \le q F^n, \tag{11}$$

where

$$F^n = \|z^n\|^2 + \tau \tau_0 \delta \sum_m \|z_{m,\bar{x}_m}^n\|^2 + \frac{\tau}{\tau_0 d_2} \|p^n\|^2, \tag{12}$$

$$q = \max \left\{ 1 - \tau \tau_0 c_0^2, \frac{1}{d_2} \right\} < 1,$$

$$d_2 = \min \left\{ 1 + \tau \nu \varepsilon_1 d_1, \ 1 + \frac{\nu \varepsilon_2}{\tau_0 \delta} \right\},$$

and c_0, d_1, ε_1, ε_2 are uniformly bounded constants that do not depend on the parameters of the grid.

It is clear from (11), (12) that the quantity q $(0 < q < 1)$ does not depend on the spatial steps of the grid, i.e. the rate of convergence of the iterative process (6)–(8) in the linear case does not depend on the number of nodes of the finite difference grid.

Next, we investigate the rate of convergence of the iterative algorithm (6)–(8) for the nonlinear case. In this case, the difference relation for the error of the solution is as follows:

$$\frac{z_m^{n+1} - z_m^n}{\tau} + L_{h,m}\left(u^n\right) z_m^{n+1} + L_{m,h}\left(z^n\right) u_m + \left(\pi^n - \tau_0 \underline{\mathrm{div}}_h z^n\right)_{x_m} =$$

$$= \nu \Delta_h z_m^{n+1} + \tau_0 \delta \left(z_{m x_m}^{n+1} - z_{m \bar{x}_m}^n\right)_{x_m}, \tag{13}$$

$$\frac{\pi^{n+1} - \pi^n}{\tau} + \underline{\mathrm{div}}_h z^{n+1} = 0.$$

It is shown for the iteration error (13) that if the data of the problem (3), (4) and the grid parameters satisfy the conditions

$$\nu - c_0 \left\|\nabla_h u\right\| (1 + \varepsilon_1) \geq \nu_0 > 0,$$

$$1 - \frac{\tau c_0 c_1}{2\varepsilon_1 h^2} \left\|\nabla u\right\| \geq \nu_1 > 0,$$

$$\delta - N \geq \delta_1 > 0,$$

then the following estimate holds:

$$E^{n+1} + \tau\tau_0 \left\|\underline{\mathrm{div}}_h z^n\right\|^2 + 2\tau\nu_0 \left\|\nabla_h z^{n+1}\right\|^2 + \nu_1 \left\|z^{n+1} - z^n\right\|^2 +$$

$$+ \tau\tau_0\delta_1 \sum_m \left\|z_{m,\bar{x}_m}^{n+1} - z_{m,x_m}^n\right\|^2 \leq E^n,$$

where

$$E^n = \left\|z^n\right\|^2 + \frac{\tau}{\tau_0}\left\|\pi^n\right\|^2 + \tau\tau_0\delta \sum_m \left\|z_{m,\bar{x}_m}^n\right\|^2$$

which guarantees the convergence of the iteration.

For comparison with other known algorithms and illustrating the possibilities of the proposed algorithm (6)–(8), we consider a problem in a cavity with a moving upper boundary for $Re = 100$ for the two-dimensional case $N = 2$. For comparison, the following implicit difference scheme was chosen:

$$\frac{u_m^{n+\frac{1}{2}} - u_m^n}{\tau} + L_h u_m^{n+\frac{1}{2}} + \overline{\mathrm{grad}}_h p^n = \nu\Delta_h u_m^{n+\frac{1}{2}} + f, \tag{14}$$

$$\frac{u^{n+1} - u^{n+\frac{1}{2}}}{\tau} + \overline{\mathrm{grad}}_h \left(p^{n+1} - p^n\right) = 0, \tag{15}$$

Table 1. Results of calculations according to scheme (14), (15)

τ	33×33	65×65	129×129
0.15	85	281	1222
0.2	88	373	1627
0.3	129	557	2438
0.35	150	650	2844

Table 2. Results of calculations according to scheme (6)–(8) for $\tau = 2$

τ_0	33×33	65×65	129×129
0.01	45	45	45
0.0125	42	43	43
0.025	42	42	43
0.03	40	43	43

$$\underline{\mathrm{div}}_h \boldsymbol{u}^{n+1} = 0.$$

Iterations were performed before the convergence criterion was fulfilled:

$$\sum_m \left\| (p^n - \tau \underline{\mathrm{div}}_h \boldsymbol{u}^n)_{x_m} - \nu \Delta_h u_m^n \right\| + \left\| \underline{\mathrm{div}}_h \boldsymbol{u}^n \right\| \leq 10^{-4}. \tag{16}$$

It should be noted that in all cases the criterion of convergence is achieved with the specified accuracy.

Tables 1 and 2 show the values of $n_0\,(\varepsilon)$, the number of iterations to satisfy the convergence criterion (16).

It can be seen from the table that the number of iterations for the algorithm (14), (15) increases when the grid step decreases.

It can be seen from the table that with an increase in the number of grid nodes, the number of iterations remains practically unchanged for the iteration scheme (6)–(8). Using the proposed algorithm, we also considered the problem of counterflows of an incompressible fluid in a channel of finite length with the boundary conditions given in Fig. 1. Numerical solutions of hydrodynamic problems for a viscous incompressible fluid use the Navier-Stokes equations written with respect to the *velocity vector, pressure* variables and in the *current function, vorticity* variables [9, 10].

Figures 2–5 show the velocity fields and isolines of the stream function for different length values of the considered region. According to the results of the obtained figures of the computational experiment, it can be concluded that in all cases the steady-state flow regime was achieved [11–14] (Table 3).

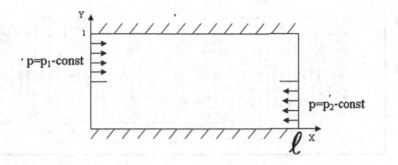

Fig. 1. Calculation region for plane flow

Table 3. Results of calculations with parameters $\tau = \tau_0 = 0.05$, $Re = 1000$

	65×65	129×129	257×257	513×513	1025×1025
$n_0\,(\varepsilon)$	368	379	381	396	419
Counting time	0:00:29	0:03:36	0:24:10	4:29:54	14:13:20

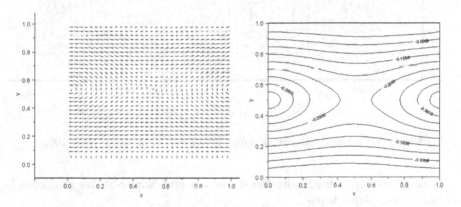

Fig. 2. The velocity field and the isoline of the current function for $Re = 500$, $l = 1$, (65×65)

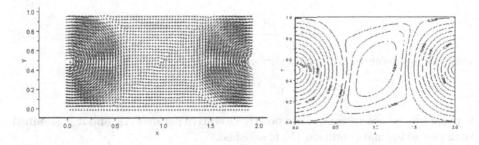

Fig. 3. The velocity field and the isoline of the current function for $l = 2$, (129×65)

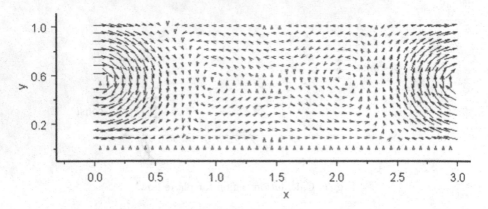

Fig. 4. Vector fields in a rectangular domain $l = 3$, (193×65), $Re = 500$

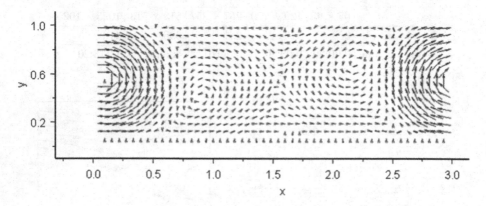

Fig. 5. Vector fields in a rectangular area $l = 3$, (193×65), $Re = 1000$

Further, to solve numerically the stationary grid Navier-Stokes problem (3), (4), a new iterative scheme

$$\left(E - \alpha \overline{\mathrm{grad}}_h \underline{\mathrm{div}}_h - \beta \Delta_h\right) \frac{u^{n+1} - u^n}{\tau} + L_h u^{n+1} + \overline{\mathrm{grad}}_h p^{n+1} = \nu \Delta_h u^{n+1} + f(x),$$
(17)

$$\frac{p^{n+1} - p^n}{\tau_0} + \underline{\mathrm{div}}_h u^{n+1} = 0$$
(18)

with homogeneous boundary conditions

$$u^{n+1}\Big|_{\partial D_h} = 0,$$
(19)

is proposed, where τ, τ_0, α, β are positive iterative parameters, and it is assumed that the additional condition (5) is satisfied.

The rate of convergence of the iteration (17)–(19) is investigated in the case of the linear Stokes problem. In this case the following theorem holds.

Theorem 2. The following estimate holds for the iterative process (17)–(19):

$$\Omega^{n+1} \leq q\Omega^n$$

where

$$\Omega^n = \|\boldsymbol{u}^n\|^2 + \alpha \|\underline{\operatorname{div}}_h \boldsymbol{u}^n\|^2 + \beta \|\nabla_h \boldsymbol{u}^n\|^2 + \frac{\tau}{\tau_0} \|p^n\|^2,$$

$$q = \left[\min\left\{ \left(1 + \frac{\tau\nu\delta}{2}\right), \left(1 + \frac{\tau\tau_0}{\alpha}\right), \left(1 + \frac{\tau\nu}{\beta}\right), (1 + c_0^2\tau\tau_0 M)^{-1} \right\} \right] < 1,$$

δ, M are positive constants. Here the number q $(0 < q < 1)$, characterizing the rate of convergence does not depend on the parameters of the spatial grid, i.e. the proposed algorithm (17)–(19) has the property of uniform convergence.

The difference relation for the error of the solution in the case of a nonlinear problem is as follows:

$$\left(E - \alpha\overline{\operatorname{grad}}_h\underline{\operatorname{div}}_h - \beta\Delta_h\right)\frac{z^{n+1} - z^n}{\tau} + L_{h,u}\left(\boldsymbol{u}^n\right)z^{n+1} + L_{h,u}\left(z^n\right)\boldsymbol{u} +$$

$$+ \overline{\operatorname{grad}}_h\pi^{n+1} = \nu\Delta_h z^{n+1} + \boldsymbol{f}\left(x\right), \tag{20}$$

$$\frac{\pi^{n+1} - \pi^n}{\tau_0} + \underline{\operatorname{div}}_h z^{n+1} = 0.$$

In this case, if we assume that

$$1 - \frac{c_0\tau\|\nabla_h \boldsymbol{u}\|}{2\varepsilon_1\beta} \geq \nu_2 > 0, \quad 1 - \frac{c_0\|\nabla_h \boldsymbol{u}\|}{\nu}\left(1 + \varepsilon_1\right) \geq \nu_1 > 0,$$

then the following theorem holds for the rate of convergence.

Theorem 3. For the error of the iterative algorithm (20), the following estimate holds:

$$E^{n+1} \leq qE^n$$

where

$$E^n = \|z^n\|^2 + \frac{\tau}{\tau_0}\|\pi^n\|^2 + \alpha\|\underline{\operatorname{div}}_h z^n\|^2 + \beta\|\nabla_h z^n\|^2,$$

$$q = \frac{1}{\min\left\{1 + \tau\delta_0\varepsilon_2, \ 1 + \tau\tau_0\gamma, \ 1 + \frac{\varepsilon_3\tau}{\beta}, \ 1 + \frac{\varepsilon_4\tau}{\alpha}\right\}} < 1,$$

and γ, ε_1, ε_2, ε_3, ε_4 are positive constants.

3   Conclusion

In this paper, the following results are obtained for the equations of an incompressible fluid in natural *velocity*, *pressure* variables:

- an implicit iterative difference scheme of the form (6)–(8) is constructed and investigated for the numerical solution of the grid stationary Navier-Stokes equations. The method of a priori estimates shows that the constructed scheme does not depend on the number of nodes in the spatial grid in the case of the linear Stokes problem, i.e. has the property of uniform convergence;
- the properties of the convergence of the iterative algorithm in a nonlinear case are characterized: it is revealed that in the case of a nonlinear problem, the study of the convergence of the iterative algorithm (6)–(8) imposes a restriction on the solution coinciding in order with the condition which guarantees the existence and uniqueness of the solution of the original differential problem;
- a new implicit multiparameter scheme of the form (17)–(19) was developed on the basis of the idea of "weak compressibility";
- an estimate of the convergence rate of the iterative algorithm (17)–(19) for linear and nonlinear stationary equations is obtained and it is proved that the proposed scheme converges to a zero stationary solution with the speed of a geometric progression.

References

1. Samarskii, A.A.: Teoriya Raznostnyh Skhem. Nauka, Moscow (1982)
2. Voevodin, A.F., Yushkova, T.V.: Chislennyj metod resheniya nachalno-kraevyh zadach dlya uravnenij nav'e - stoksa v zamknutyh oblastyah na osnove metoda rasscheplenii. Sibirskij Zhurnal Vychislitelnoj Matematiki **2**, 321–332 (1999)
3. Zhilin, L., Wang, C.: A fast finite differenc method for solving navier-stokes equations on irregular domains. Commun. Math. Sci. **1**, 180–196 (2003)
4. Papin, A.A.: Razreshimost "v malom" ponachalnym dannym uravnenij odnomernogo dvizheniya dvuh vzaimopronikayuschih vyazkih neszhimaemyh zhidkostej. Dinamika sploshnoj sredy **116**, 73–81 (2000)
5. Zhumagulov, B.T., Temirbekov, N.M., Baigereyev, D.R.: Efficient difference schemes for the three-phase non-isothermal flow problem. In: AIP Conference Proceedings, vol. 1880, pp. 060001:1–060001:10 (2017)
6. Wójcik, W., Temirbekov, N.M., Baigereyev, D.R.: Fractional flow formulation for three-phase non-isothermal flow in porous media. Przegl. Elektrotech. **92**(7), 24–31 (2016)
7. Danaev, N.T., Yergaliev, Y.K.: Ob odnom iteratsionnom metode resheniya statsionarnyh uravnenii Nav'e-Stoksa. Vychislitel'nye Tekhnologii **11**, 37–43 (2006)
8. Danaev, N.T., Yergaliev, Y.K.: Ob odnoj neyavnoj iteratsionnoj skheme dlya zadachi Stoksa. Vestnik KazNU. Seriya matematika, mekhanika i informatika 50 (2006)
9. Danaev, N.T., Amenova, F.S.: About one method to solve Navier-Stokes equation in variables (Ω, Ψ). Adv. Math. Comput. Math. **3**, 72–78 (2013)
10. Dajkovskij, A.G., Polezhaev, V.I., Fedoseev, A.I.: O raschete granichnyh uslovij dlya nestatsionarnyh uravnenij Nav'e-Stoksa v peremennyh (ψ, ω). Chislennye Metody Mekhaniki Sploshnoj Sredy **10**, 49–58 (1979)
11. Merzlikina, D.A., Pyshnograj, G.V., Pivokonskij, R., Filip, P.: Reologicheskaya model' dlya opisaniya viskozimetricheskih techenij rasplavov razvetvlennyh polimerov. Inzhenerno-fizicheskij zhurnal **89**, 643–651 (2016)

12. Merzlikina, D.A., Pyshnograi, G.V., Koshelev, K.B., Kuznetcov, A., Pyshnograi, I.G., Tolstykh, M.U.: Mesoscopic rhelogical model for polymeric media flows. J. Phys. Conf. Ser. **1**, 195–197 (2017)
13. Temirbekov, A.N., Urmashev, B.A., Gromaszek, K.: Investigation of the stability and convergence of difference schemes for the three-dimensional equations of the atmospheric boundary layer. Int. J. Electron. Telecommun. **64**, 391–396 (2018)
14. Wojcik, W., et al.: Probabilistic and statistical modelling of the harmful transport impurities in the atmosphere from motor vehicles. Rocznik Ochrona Srodowiska **19**, 795–808 (2017)

Mathematical Modeling of Gas Generation in Underground Gas Generator

Y. N. Zakharov[✉]

ICT of SB RAS, Kemerovo State University,
6, Krasnaya Street, Kemerovo, Kemerovo Region 650043, Russia
`zaxarovyn@rambler.ru`, `zyn@kemsu.ru`

Abstract. Underground coal gasification is an in-situ underground physical and chemical process which converts coal into combustible gases using injections of free or bound oxygen. The paper presents two dimensional nonstationary mathematical model of underground gas generator performance based on flux method. It takes into account physical, chemical and gas dynamic processes. Gas medium consists of the following gases: CH_4, H_2, CO, O_2, H_2O, N_2. The research shows the numerical calculations of gas composition change inside the gas generator describing table values of brown and bituminous coal combustion products. *abstract* environment.

Keywords: Underground coal gasification · Compressible flow · Difference methods · Lateral fire well · Combustion face

1 Introduction

Advanced coal processing is one of the most complex and time consuming tasks of coal industry. This task solution can improve economic performances of fuel and energy industry and deal with the issues of ecological safety. Underground coal gasification (UCG) can be one of these problem solutions. Underground coal gasification is an in-situ underground physical and chemical process which converts coal into combustible gases using injections of free or bound oxygen. Establishing UCG production is possible for those places where deep mining and surface mining are not commercially viable. Underground coal gasification enables to dig for coal in the context of flat-lying high ash coal seams. One of the UCG advantages is that there are no severe surface damages caused. Moreover gas is considered to be environmentally friendly fuel: processed UCG gas has no hydrogen sulfide and does not release sulfur dioxide during combustion. The gas produced with oxygen injection has no nitrogen oxides [1]. UCG production has significant health and safety advantages: no people required to work underground, no work accidents associated with deep mining.

The study was conducted with the financial support of the internal grant of the Kemerovo University.

Y. Shokin and Z. Shaimardanov (Eds.): CITech 2018, CCIS 998, pp. 218–227, 2019.
https://doi.org/10.1007/978-3-030-12203-4_22

Underground coal gasification technology was developed by the Soviet engineers in the 30's. Some experimental facilities were built in the USSR in the 50's. Yuzhno-Abinskaya station "Podzemgas" in Kuzbass (1955–1996) is one of them. Underground coal gasification seemed to be of great interest for foreign countries in the 70's and 80's. Major coal mining countries paid for UCG technology licenses. Though early UCG experiments naturally took place 80 years ago, there have been no industrial facilities until quite recently because of the complexity of UCG process technology, depending on many factors (mine engineering, hydrodynamic, hydro geologic, etc).

Nowadays scientists identify possible hydrocarbon energy potential of the Earth as 100% that includes oil (4.2%), gas (2.5%) and coal (93.3%) [2]. Steady increase of coal importance for the future economy is registered not only in Russia but in the whole world as well and makes unconventional coal mining technology (UCG) be the issue of great interest.

A great interest for UCG commercialization has been observed abroad during recent years. China and Australia are demonstrative example of it. The article [3] lists Chinese coal companies that are currently engaged in UCG implementation, e.g.: Xinwen Coal Industry Group; Feicheng Coal Industry Group; Xiyang Chemistry Company, etc. Now Australia has taken top position in the field of UCG. There are group of the companies in Australia that succeed in UCG exploitation. Here is the list of the most known and largest companies in the world: "Linc Energy" (global UCG leader) and "Australian Syngas Association Inc" (represents the group of Australian UCG companies). Currently UCG technology has small amount of theoretical backgrounds. Though UCG technology is widely used today, though development of relevant mathematical model and its validation is still considered to be a crucial task despite the researches [4–8].

Most researches contain simplified models that are basis for engineering formulae for received gas content calculation. The literature reviewing UCG issues also contains researches devoted to the more extensive description of physical and chemical transformations in the reaction site. In the same time there are no data on the complete composition of the received gas, and there is no comparison with the results of in situ tests. Thus, the task of validation of the reviewed UCG model has not been solved. Published in 2004 E. V. Kreinin's research contains much information comparing mathematical simulation based on engineering methods and in situ measurements taken at the operating flow method UCG companies.

Most researches contain simplified models that are basis for engineering formulae for received gas content calculation. The literature reviewing UCG issues also contains researches devoted to the more extensive description of physical and chemical transformations in the reaction site [9]. In the same time there are no data on the complete composition of the [8] received gas, and there is no comparison with the results of in situ tests. Thus, the task of validation of the reviewed UCG model has not been solved. Published in 2004 E. V. Kreinin's research contains much information comparing mathematical simulation based on engineering methods and in situ measurements taken at the operating flow method UCG companies.

Consequently, computational modeling of gas generation based on flux method and UCG mathematical model validation (described in [10]) are the key goals of the paper. Gas composition determined at the exit of the gas production well (with the help of numerical calculations) is compared to the results of in-situ measurements of gas composition changes.

2   UCG Model

The paragraph is devoted to the UCG model based on flux method [8] (1).

To install gas generator, coal seam with at least 5 m thickness and 30–800 m formation depth is needed.

Two wells are drilled on either side of an underground coal seam, a lateral well (so called "lateral fire well") is drilled to connect the two vertical wells. It is used to ignite and fuel the underground combustion process, so the coal face ("combustion face") burns. One well is used to inject air or steam-oxygen (injection well) into the coal seam. The second well (gas collecting well) is used to collect the gas that is formed from the gasification reactions and to pipe it to the surface. As the coal face burns, the immediate area is depleted from the bottom up. Burning front shifts in the same direction. The remaining cavity usually contains the left over ash and fallen parts of the roof. The lateral fire well section barely changes due to coal burning, and the burning face surface remains available for injected blast, and as shown in [8] the gasifier operation is being stabilized. The injected blast flows round combustion face surface, gasifies coal and causes combustion gas generation. Both burning process and gasification process is considered to be single process. Part of the heat formed from the combustion process is transmitted to the immediate coal area. Heating process contains two stages. Firstly, coal moisture evaporates (drying of coal), the process decomposes coal and generates combustible volatiles (mostly CH_4, H_2) and carbon residue that contains carbon and ash. Coal temperature increases. Afterwards, coking residue carbon heterogeneously reacts with free and bound oxygen and water vapor to transform into CO and incombustible gases. The temperature can be 1500K–1700K. Remaining part of the heat is used to heat up gases in the lateral fire well. Gases generated during the decomposition reaction and further oxidation of coke carbon can be divided into two portions. Some gas portion is filtered and gets into unmined coal due to pressure gradient. The remaining gas portion gets into lateral fire well to heat up gas mixture and cause homogeneous oxidizing reactions due to convection and diffusion. The research [10] shows mathematical model of UCG that takes into consideration the physical processes of coal gasification mentioned above, gas flow in lateral fire well and mine face form change.

Two dimensional mathematical model of UCG [10] that describes UCG processes taking place both in unmined coal and lateral fire well is under consideration. As far as gas quality basically depends on the processes taking place on the combustion face surface and in the lateral fire well, so when computational modeling is concerned, the quotation describing UCG only in gasifier channel Ω_2 and at combustion face (Γ_1) is under consideration in this paper (Fig. 1). Solution of a complete model including gas seepage, ash fall and fire face move demands another article layout. Here we are not concerned by mass values of gas seepage, here it is necessary to determine the content of the receiving gas and compare it with the gas content received by UCG from different coal basins. According to the mathematical model [6] the gas in lateral fire well is compressible and viscous and consists of seven components: CH_4, H_2, CO, O_2, H_2O, N_2

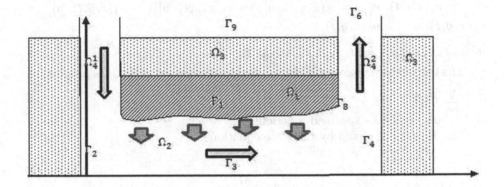

Fig. 1. Underground coal gasification (UCG)

Here are the main symbols used:
Ω_1 - coal seam, Ω_2 - lateral fire well, $\Omega_4^1 \cup \Omega_4^2$ - injection well and gas collection well, Ω_3 - soil, Γ_1 - combustion face, Γ_2, Γ_7 - side boundaries of the gas collecting well, Γ_4, Γ_8 - side boundaries of the gas collection well, Γ_3 - bottom interface of the lateral fire well, Γ_5, Γ_6 - entry and outlet section of the wells, Γ_9 - ground surface, $\rho(x, y, z)$ and $p(x, y, z)$ - density and pressure of the gas mixture, $u(x, y, z)$, $v(x, y, z)$ - projection of the gas velocity vector on the axis x and y relatively; $T(x, y, z)$ - temperature of the gas mixture, μ, μ_k - viscosity of the gas mixture and gas k-element, M, M_k - molar mass of the k-element mixture, λ, λ_k and c_p, c_{p_k} - thermal conduction and specific thermal capacity of the gas mixture and gas k-element, D_k - effective diffusion factor of the k- gas, c_k - proportion of k-element of the gas mixture, R - universal gas constant; q_k and P_k - enthalpy of formation, C_k^{out} and T_{out} - fractional gas composition and exterior temperature, α - heat-transfer coefficient, k_j, E_j, q_j - pre-exponential factor, activation energy and thermal effect of homogeneous reactions, W_k - mass change rate of gas phase k-component, R_j - mass rates of heterogeneous reactions, M_y - molar mass of dry coal, v_i, v_{3k} - stoichiometric coefficients, s - pore surface per unit volume of porous medium, ω_i - volume ratio of porous medium i-phase, where $i = 1$ - dry coal, $i = 2$ - moisture, $i = 3$ - gas phase, $i = 4$ - charred coal, $i = 5$ - ash, $f(x)$ - combustion face form, n - normal vector to the boundary.

Gaseous phase is considered to be a combination of non-viscous perfect gases, while its component diffusion is independent. The flow in the fields is laminar.

In $\Omega_2 \cup \Omega_4^1 \cup \Omega_4^2$ (Fig. 1) viscous compressible heat-conducting gas flow is defined by the Navier-Stokes equation system that describes nonstationary flow of viscous heterogeneous compressible fluid:

$$\rho \left(\frac{\partial u}{\partial t} + u\frac{\partial u}{\partial x} + v\frac{\partial u}{\partial y} \right) = -\frac{\partial p}{\partial x} + \frac{\partial}{\partial x}\left(\mu\frac{\partial u}{\partial x} \right) + \frac{\partial}{\partial y}\left(\mu\frac{\partial u}{\partial y} \right),$$
$$\rho \left(\frac{\partial v}{\partial t} + u\frac{\partial v}{\partial x} + v\frac{\partial v}{\partial y} \right) = -\frac{\partial p}{\partial y} + \frac{\partial}{\partial x}\left(\mu\frac{\partial v}{\partial x} \right) + \frac{\partial}{\partial y}\left(\mu\frac{\partial v}{\partial y} \right),$$
$$\tag{1}$$

Here are the initial: $v_0(x,y,0) = u_0(x,y,0)$, and boundary conditions:
$v|_{\Gamma_1} = v_1(x,y,t)$, $v|_{\Gamma_5} = v_5(x,y,t)$, $v|_{\Gamma_6} = v_6(x,y,t)$, $u|_{\Gamma_1} = u_1(x,y,t)$, $u|_{\Gamma_5} = u_5(x,y,t)$, $u|_{\Gamma_6} = u_6(x,y,t)$.
$\frac{\partial v}{\partial n}\big|_{\Gamma_2,\Gamma_3,\Gamma_4,\Gamma_7,\Gamma_8} = \frac{\partial u}{\partial n}\big|_{\Gamma_1,\Gamma_2,\Gamma_3,\Gamma_4,\Gamma_5,\Gamma_6,\Gamma_7,\Gamma_8} = 0.$

The gas mixture viscosity is determined in accordance with k-gas proportion.
$\mu = \sum\limits_{k=1}^{7} \mu_k c_k$
and v_0, u_0, v_1, v_5, v_6 - specified functions.

The Eq. (1) are closed by the state equation

$$p = \frac{\rho R T}{M}, \tag{2}$$

and molar mass of the gas phase is calculated in the following way
$\frac{1}{M} = \sum\limits_{k=1}^{7} \frac{c_k}{M_k}$

The following heat-transfer equation governs heat transfer, absorption and emission in the gasifier:

$$\rho c_p \left(\frac{\partial T}{\partial t} + u\frac{\partial T}{\partial x} + v\frac{\partial T}{\partial y} \right) = \frac{\partial}{\partial x}\left(\lambda\frac{\partial T}{\partial x} \right) + \frac{\partial}{\partial y}\left(\lambda\frac{\partial T}{\partial y} \right) + \sum\limits_{k=3}^{6} q_k P_k \tag{3}$$

taking into consideration the initial: $T_0 = T_0(x,y,0)$ and boundary conditions:
$T|_{\Gamma_1} = T_1(x,y,t)$, $T|_{\Gamma_5} = T_5(x,y,t)$, $\frac{\partial T}{\partial n}\big|_{\Gamma_2,\Gamma_3,\Gamma_4,\Gamma_7,\Gamma_8} = 0$, $\frac{\partial T}{\partial n}\big|_{\Gamma_6} = \alpha(T - T_{out})$,
where α - heat-transfer coefficient, T_0, T_1, T_5 - specified functions of initial temperature and temperatures on the boundaries Γ_1 and Γ_5.

Mass rates of the reactions P_k, $k = 3,\ldots,6$ are calculated by using the following formulas

$$P_3 = k_3 \rho c_1 e^{-\frac{E_3}{RT}}, P_4 = k_4 \rho c_2 e^{-\frac{E_4}{RT}},$$
$$P_5 = k_5 \rho c_3 e^{-\frac{E_5}{RT}}, P_6 = k_6 \rho c_4 e^{-\frac{E_6}{RT}},$$

where k_j, E_j, q_j - thermokinetic constants, λ and c_p coefficients are calculated for the gas mixture taking into consideration proportion of each component

$$\lambda = \sum_{k=1}^{7} \lambda_k c_k, \qquad c_p = \sum_{k=1}^{7} c_k c_{pk}.$$

Equation of continuity is relevant for the gas in gasifier channel

$$\frac{\partial \rho}{\partial t} + \frac{\partial}{\partial x}(\rho u) + \frac{\partial}{\partial y}(\rho v) = 0 \tag{4}$$

with initial data $\rho_0 = \rho_0(x, y, 0)$ where ρ_0 is a specified function, $(x, y) \in \Omega_2 \cup \Omega_4^1 \cup \Omega_4^2$.

As long as gas phase contains seven components, and each component has its own physical specifications, diffusion and transfer process is considered to be determined by individual convection-and-diffusion equation for each component:

$$\rho \left(\frac{\partial c_k}{\partial t} + u \frac{\partial c_k}{\partial x} + v \frac{\partial c_k}{\partial y} \right) = \frac{\partial}{\partial x}\left(\rho D_k \frac{\partial c_k}{\partial x} \right) + \frac{\partial}{\partial y}\left(\rho D_k \frac{\partial c_k}{\partial y} \right) + W_k, \, k = 1, ..., 6$$
$$\sum_{k=1}^{7} c_k = 1. \tag{5}$$

with initial conditions

$$c_1(x, y, 0) = c_1^0, c_2(x, y, 0) = c_2^0, c_3(x, y, 0) = c_3^0, c_4(x, y, 0) = c_4^0,$$
$$c_5(x, y, 0) = c_5^0, c_6(x, y, 0) = c_6^0, c_7(x, y, 0) = c_7^0,$$

and boundary conditions

$$c_k|_{\Gamma_1} = c_{k1}(x, y, 0), c_k|_{\Gamma_5} = c_{k5}(x, y, 0), k = 1..6$$
$$\frac{\partial c_k}{\partial y}\Big|_{\Gamma_6} = \beta(c_k - c_k^{out}), k = 1..6$$
$$\frac{\partial c_k}{\partial n}\Big|_{\Gamma_2, \Gamma_3, \Gamma_4, \Gamma_7, \Gamma_8} = 0, k = 1..6,$$

where $c_{k1}, c_{k5}, k = 1..6$ - specified functions of the proportions of the gas mixture k-element on the boundaries Γ_1 and Γ_5.

In this case, the formulas for W_k demand $P_1 = P_2 = s = 0$, $\varphi_3 = 1$ as Ω_2 is gas flow region.

For example, for methane $CH_4 : W_1 = \frac{\nu_{31}}{\nu_1} \frac{M_1}{M_y} P_1 - \varphi_3 P_3,$

For hydrogen $H_2 : W_2 = \frac{\nu_{32}}{\nu_1} \frac{M_2}{M_y} P_1 + \varphi_3 (\frac{M_2}{M_3} P_6 - P_4) + s \frac{M_2}{M_c} R_4 = \varphi_3 (\frac{M_2}{M_3} P_6 - P_4).$

The following heterogeneous chemical reactions of carbon oxidizing and carbon monoxide reduction can take place during UCG process

$$(1)\, C + O_2 = CO_2 + q_{s1}, \qquad (2)\, C + \tfrac{1}{2}O_2 = CO + q_{s2},$$
$$(3)\, C + CO_2 = 2CO + q_{s3}, \qquad (4)\, C + H2O = CO + H_2 + q_{s4},$$

taking into consideration corresponding absolute value of mass rates:

$$R_1 = \frac{M_c}{M_4} k_{s1} \rho_3 c_4 e^{-\frac{E_{s1}}{RT}}, \quad R_2 = \frac{M_c}{M_4} k_{s2} \rho_3 c_4 e^{-\frac{E_{s2}}{RT}}$$
$$R_3 = \frac{M_c}{M_4} k_{s1} \rho_3 c_5 e^{-\frac{E_{s3}}{RT}}, \quad R_4 = \frac{M_c}{M_4} k_{s2} \rho_3 c_6 e^{-\frac{E_{s3}}{RT}} \tag{6}$$

Absolute values of mass rates of pyrolysis reaction - P_1 and moisture evaporation - P_2 are determined by the Arrhenius law and simplified Hertz-Knudsen law

$$P_i = k_i \rho_i \phi_i e^{-\frac{E_i}{RT}}, i = 1, 2, .$$

Mass rates of the W_k gas phase k-component changes are determined in the following way

$CH_4 : W_1 = \frac{v_{31}}{v_1} \frac{M_1}{M_y} P_1 - \phi_3 P_3,$

$H_2 : W_2 = \frac{v_{32}}{v_1} \frac{M_2}{M_y} P_1 + \phi_3 \left(\frac{M_2}{M_3} P_6 - P_4 \right) + s \frac{M_2}{M_c} R_4,$

$CO : W_3 = \frac{v_{33}}{v_1} \frac{M_3}{M_y} P_1 - \phi_3 (P_5 + P_6) + s \frac{M_3}{M_c} (R_2 + 2R_3 + R_4),$

$O_2 : W_4 = -\phi_3 \left(2\frac{M_4}{M_1} P_3 + \frac{1}{2}\frac{M_4}{M_2} P_4 + \frac{1}{2}\frac{M_4}{M_3} P_5 \right) - s\frac{M_4}{M_c} \left(R_1 + \frac{1}{2} R_2 \right),$

$CO_2 : W_5 = \frac{v_{35}}{v_1} \frac{M_5}{M_y} P_1 + \phi_3 \left(\frac{M_5}{M_1} P_3 + \frac{M_5}{M_3} P_5 + \frac{M_5}{M_3} P_6 \right) + s\frac{M_5}{M_c} (R_1 - R_3),$

$H_2O : W_6 = \frac{v_{36}}{v_1} \frac{M_6}{M_y} P_1 + P_2 + \phi_3 \left(2\frac{M_6}{M_1} P_3 + \frac{M_6}{M_3} P_4 - \frac{M_6}{M_3} P_6 \right) - s\frac{M_6}{M_c} R_4,$

$N_2 : W_7 = 0.$

Combustion face form $f(x, t)$ is identified as the solution of nonlinear equation [8]

$$\rho_4 \frac{\partial f}{\partial t} - \sqrt{1 + \left(\frac{\partial f}{\partial x} \right)^2} \cdot \sum_{j=1}^{4} R_j = 0 \qquad (7)$$

with initial data $f|_{t=0} = f^0(x)$, where $f^0(x)$ - specified initial form of the combustion face, and the function R_j is calculated by the formula (6).

Consequently, the equation system (1)–(6) with corresponding initial and boundary conditions describes the processes in lateral fire well and combustion face. Numerical algorithm of the UCG problem is determined in accordance with the presented mathematical model. Firstly, the Cauchy problem is solved for the continuity equation (4) by using known values of velocities and specified initial data. Secondly, temperature propagation equations are solved (3). Thirdly, Navier-Stokes motion equations are solved after calculating the pressure value determined by state equation (2). As soon as all necessary flow state changes are determined at a new time step, changes of gas composition are calculated (5) and new state of mine face form is identified (7).

3   Results of the Numerical Experiments

This section is devoted to the results of the numerical experiments of nondimensionalized UCG mathematical model (1)–(6) (black coal and brown coal).

Numerical experiments are carried out in full accordance with common methods: calculations take into consideration progressively fine meshes and solutions' comparison, robustness test of numerical methods, solutions of the problems depending on various initial data, etc.

Unequally spaced mesh (as far as space variables are concerned) Ω_h with steps $h_{x_{ij}}$, $h_{y_{ij}}$ and constant time-step $\tau > 0$ is considered to be in the domains

$\Omega = \Omega_1 \cup \Omega_2 \cup \Omega_3 \cup \Omega_4^1 \cup \Omega_4^2$. N and M define number of points on the axis OX and OY respectively.

The Eqs. (1), (3)–(5) are approximated on Ω_h in a standard way [11] by using difference schemes. To solve continuity Eq. (4) the Lax-Wendroff-type scheme [11] with implicit viscosity is used. The system (1) and the Eq. (3), (5) are solved with the help of difference scheme of stabilizing correction with directional differences [12]. Viscosity coefficient μ as well as other coefficients use fractional gas composition of the previous time moment to solve all the equations. The equation of mine face form change is solved by the first approximation order scheme in the context of time and space.

Coefficient values μ_k, λ_k, c_{pk} for k-gas are mentioned in [13–15], and thermokinetic constants q_{3-6}, k_{3-6}, E_{3-6}, M_{3-6}, M_y, M_c are mentioned in [16–18]. Brown coal ash content is considered to be 30% that is similar to coal ash content in Moscow lignite basin [19]. Field observation results (mentioned in [20]) are considered to be input data for gas composition that is caused by coal thermal decomposition.

Comparison on a percentage base of rated gas composition to gas compositions described in different researches is presented further. Table 1 shows percentage composition of the gas calculated for uniform gasification process of bituminous coal and the in-situ measurements of gas composition, which are relevant for different coal-bearing basins [8,21]. Table 1 shows the intervals that limit test values of gas composition. The researches [8,21] show in-situ measurements (in the context of air blast) taken in Kuzbass Yuzhno-Abinskaya coalmine "Podzemgas". The column No3 shows real measurements taken by E.V. Kreinin, who is considered to be one of the principal UCG researchers and to spend much time on its analysis and development. He presents UCG engineering model in his paper [8]. His natural experiments are based on that model.

Table 1. The percentage of gas in the gasification of coal.

	1.	2.	3.
	Rated gas composition according to the model (1)–(7) %	Real gas composition (Yuzhno-Abinskaya station) [21] %	Gas composition according to Kreinin [8] %
CH_4	1.9	1.6–3.	2.6
H_2	10.6	10–15	12.5
CO	17.2	10–20	11.9
O_2	0.2	0.2	0.2
CO_2	10.9	8.0–14.5	13.2
N_2	59.1	53–63	59.5
Nonregistered impurity	0.1	0.1–0.5 0.01–0.02	0.1

Table 2 shows the comparison results of numerical calculations to in-situ measurements of brown coal. The research [20] presents a range of gas composition values taken in Moscow lignite basin and at Shatskay station. The table columns 2–3 show that different types of gas can be generated (in the context of brown coal) because of various coal characteristics in different basins. Calculation results based on laboratory data concerning coal decomposition show true burning processes registered in brown coal of different deposits.

Table 2. The percentage of gas in the gasification of drilling coal.

	1.	2.	3.
	Rated gas composition according to the model (1)–(7) %	Real gas composition (Podmoskovnaya and Shatskaya) [21] %	Gas composition according to Kreinin [8] %
CH_4	1.97	1.0–1.5	2.0
H_2	14.8	15–17	22.5
CO	9.7	5–7	4
O_2	0.2	0.3–0.5	0.4
CO_2	18.9	17–18	21.5
N_2	53.2	56–59	49.0
Nonregistered impurity	1.23	–	0.2

The presented calculations show that the underground gasifier model described is able to perform real UCG processes, which take place in both bituminous and brown coal mines and enables to get valid quantitative agreement with in-situ measurements.

The presented calculations show that the underground gasifier model described taking account of physical and chemical transformations and gasdynamics of thermally conductive compressible gas is able to perform real UCG processes flow method, which take place in both bituminous and brown coal mines and enables to get valid quantitative agreement with field measurements. Thus, the validation of the mathematical model carried out will allow the conducted experiments to determine the input parameters for the optimal operation of the gas generator.

References

1. Bodnaruk, M.N.: Ecological and economic analysis of modern underground coal gasification situation in coal mines. Inf. Anal. Min. Bull. **6**, 272–273 (2011)
2. Zakharov, E.I., Kachurin, N.M.: Coal and its role in development of Russian regions. Tula State University news. Nauki o zemle 1 (2014)
3. Lazarenko, S.N., Kravtsov, P.V.: A new stage in the development of underground gasification of coal in Russia and in the world. Min. Inf. Anal. Bull. Econ. Econ. **5**, 304–316 (2007)
4. Kotsur, K., Kacur, J.: The mathematical modeling of relevant processes for underground coal gasification. In: Proceeding of 5th International Symposium on Earth Science and Technology, University Fukuoka, Fukuoka, pp. 475–480 (2007)
5. Uppala, A.A., Bhattib, A.I., Aamirb, E., Samarb, R., Khana, S.A.: Control oriented modeling and optimization of one dimensional packed bed model of underground coal gasification. J. Process Control **24**(1), 269–277 (2014)
6. Aghalayam, P.: Mathematical Modeling for Underground Coal Gasification, p. 68. Momentum Press, New York (2017)
7. Sundararajan, T., Raghavan, V., Ajilkumar, A., Vijay, K.K.: Mathematical modelling of coal gasification processes. In: UCG-2017 IOP Publishing IOP Conference Series: Earth and Environmental Science, vol. 76, p. 012006 (2017)
8. Kreinin, E.V.: Unconventional Thermal Technologies of Hard-to-Recover Fuel Extraction: Coal, Crude Hydrocarbons. NRC Gasprom, Moscow (2004)
9. Subbotin, A.S., Kulesh, R.N., Mazanik, Λ.S.: Mathematical modeling of heat and mass transfer at underground coal gasification. Bulletin of the Tomsk Polytechnic University, vol. 325, no. 4 (2014)
10. Zakharov, Y.N., Zelenskii, E.E., Shokin, Y.I.: A model of an underground gas problem. RJNAMM **28**(3), 301–317 (2013)
11. Roache, P.: Computational Fluid Dynamics. Mir, Moscow (1980)
12. Yanenko, N.N.: Fractional Step Method of Mathematical Physics Multidimensional Problems Solving. Nauka, Novosibirsk (1967)
13. Kikoin, I.K.: Physical quantity tables. Reference guide. Edited by member of the academy. Atomizdat, Moscow (1976)
14. New reference guide for chemist and process engineer. General information about materials. Physical characteristics of most important materials. Laboratory research technology. Intellectual property. Saint Petersburg, Mir i Semya (2006)
15. Engineer reference guide: [Digital resource] 2006–2012 (2012). http://www.dpva.info. Accessed 10 Jan 2012
16. Alekseev, B.V.: Physical gas dynamics of reactive mediums: manual for mathematical and physical institutes. Moscow, Vyshaya shkola (1985)
17. Grishin, A.M.: Mathematical models of woods fire and new fire fighting methods. Nauka, Novosibirsk (1992)
18. Pomerantsev, V.V., Arefiev, K.M., Akhmedov, D.B.: Fundamental concepts of combustion theory: manual for universities, 2nd edn. updated and revised. Leningrad, Energoatomizdat, Leningrad branch (1992)
19. Moscow lignite basin. Endorsed by V.A. Potapenko. Tula, Grif and Co. (2000)
20. Knorre, G.F.: Burning processes. Gosenergoizdat, Moscow-Leningrad (1959)
21. Porokhov, A.M.: Chemical encyclopedia. 2 V. Moscow, Soviet encyclopedia, vol. 2 (1988)

Finite-Element Method in Tasks of Loose Soil Erosion

Y. N. Zakharov[1,2], K. S. Ivanov[1,2(✉)], and I. E. Saltykov[1,2]

[1] Kemerovo State University, Kemerovo, Russia
[2] KD of Institute of Computational Technologies of the SB RAS, Kemerovo, Russia
topspin83@mail.ru

Abstract. Finite-element method is applied to numerical simulation of loose soil erosion process. Adapted triangle grid is used to provide agreement with the time-varying mud line. The test cases are presented for some parameters of soil erosion model such as density and particle diameter. The results of computation demonstrate the vital difference between algorithms without restructuring of the grid and in case of applying it.

Keywords: Viscous incompressible fluid · Navier-Stokes equations · Non-cohesive soil erosion · Three-dimensional flow · Gravity type oil platforms · Numerical and laboratory-based experiments · Finite-element methods

1 Introduction

The application of gravity type oil platforms at shallow water marine coastal areas is one of the most current means for oil extraction. Processes of sea floor erosion near foundations of such oil platforms and its stability issues are of great interest. In the recent years different investigations of those issues were actively undertaken, by means of laboratory-based and seminatural experiments as well as by means of mathematical simulation [1–4].

The papers [5–7] contain results of a great number of experimental and numerical studies of non-cohesive soil erosion near the foundation of the Prirazlomnaya platform, comparison charts of laboratory and simulation experiments, analysis of the impact of different wave conditions of fluid flow on the process of particles shift of seabed material. Turbulence mode influence on the process of non-cohesive soil erosion was studied in parer [8].

For numerical computation of hydrodynamic values in the indicated papers they applied the models based on finite difference and finite volume methods where the fixed grid constructed at the initial instant with the condition of bottom surface evenness was used for calculation of fluid flow. First and foremost, it was stipulated by the assumption of immaterial effect of the change of the bottom shape on the fluid flow pattern according to supposedly small levels of scours and accretions. In fact, the results of the study [7,8] show that under the condition of low velocity of fluid flow in case of the nonoccurrence of surface waves

Y. Shokin and Z. Shaimardanov (Eds.): CITech 2018, CCIS 998, pp. 228–235, 2019.
https://doi.org/10.1007/978-3-030-12203-4_23

even the laminar model of fluid flow provides good agreement with laboratory tests (accurate to 10–15%). In case of surface waves and high velocity neither laminar nor turbulent flow models used in the papers [7,8] give any satisfactory results compared with laboratory experiments. One of apparent premises for disagreement of computational and laboratory studies under the specified conditions would be a significant effect of the shape of bottom on fluid flow pattern that demands the application of a time-depended grid for discretization of computational domain.

This paper develops computational models for studying loose soil erosion within the framework of the designed computing system "XFlow" [9] with the use of finite-element approximations providing agreement of the computational grid with the time-varying mud line.

2   Fluid Flow Model

Unsteady flow of viscous incompressible fluid with fixed properties in dimensionless form is described by the system of Navier-Stokes equations:

$$\begin{cases} \nabla \cdot \overline{U} = 0, \\ \frac{\partial \overline{U}}{\partial t} + (\overline{U} \cdot \nabla)\overline{U} = -\frac{1}{\rho}\nabla p + \nabla\left(\nu\nabla\overline{U}\right), \end{cases} \tag{1}$$

where $\overline{U}$ - velocity vector, t - time, ρ - fluid density, p - pressure, ν - kinematic viscosity of fluid.

Physical factor split scheme [10] is applied for time numerical integration:

$$\begin{array}{l} \frac{\tilde{U}-U^n}{\tau} + (\tilde{U}^n \cdot \nabla)\tilde{U} = \frac{1}{Re} \triangle \tilde{U}, \\ \triangle p^{n+1} = \frac{1}{\tau}(\nabla \cdot \tilde{U}), \\ \frac{U^{n+1}-\tilde{U}}{\tau} = -\frac{1}{\rho}\nabla p^{n+1}. \end{array} \tag{2}$$

In this paper the finite-element method [11] is applied for scheme (2) spatial discretization that results in use of a numerical algorithm at each time interval for sequential solution of system of linear algebraic equations (SLAE) respective to unknown variables. The stabilized biconjugate gradient method is applied for the first stage of the scheme (2) considering asymmetry of SLAE matrix. The conjugate gradient method is applied for the second stage of the scheme (2) as SLAE matrix is symmetrical. At the third stage of the scheme (2) SLAE matrix represents mass matrix [12] which admits diagonalization, hereby allowing to find the solution by direct conversion.

In order to test the applied numerical algorithms for computation of fluid flow we undertook the series of computations of known model tasks under laminar conditions [13], such as two-dimensional flow around a circular cylinder, three-dimensional flow in a driven cavity, flow over backward-facing step, flow around a cube, and flow around a sphere. The results of test computations agree rather well with analogous computations undertaken by other authors [14].

Also, we carried out the computation of task on flow near a gravity type platform considered in the papers [7,8]. The comparative analysis of quantitative properties of seabed layer flow has shown good agreement of computation results with analogous results in [7,8].

3 Soil Erosion Model

In order to compute soil erosion, the present paper applies a model considered in [15,16] and used in [5–8] for numerical simulation of soil erosion near gravity type oil platforms coastal sea areas. The overall equation of this model is mass balance equation

$$\frac{\partial h}{\partial t} = \frac{1}{1 - \epsilon} \cdot \sum_i \frac{\partial q_i}{\partial x_i}, \quad i = 1,\, 2, \tag{3}$$

written with regard to function h of elevation of the developed bottom profile above its initial level; here ϵ – porosity of sea floor material, q – vector of sea floor material transport in time unit by length unit.

The present paper stipulates the computation of a new boundary of computational region at each time interval on the basis of received h values and the restructuring of finite-element grid (Figs. 1 and 2). The obtained grid is transferred to the numerical model of fluid flow for computing a new layer, thus providing the feedback path between soil erosion profile and fluid flow.

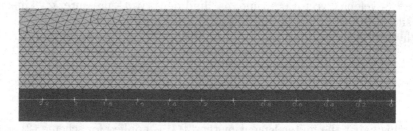

Fig. 1. Discretization of computational region: initial grid.

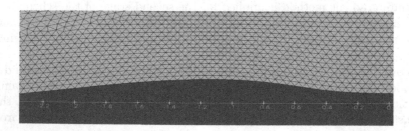

Fig. 2. Discretization of computational region: deformed grid.

4   Computational Study Results

The summarized numerical algorithm consists of three stages executed at each discrete time interval: computation of fluid flow and obtaining of velocity vector values in sea floor layers; computation of transport vector of sea floor material and elevation profile; computation of a new boundary of computational region and restructuring of finite-element grid. In order to test this algorithm we undertook the series of computations of two-dimensional task on soil erosion due to the flow around a rectangular pile under the laminar conditions of fluid flow.

Figures 3 and 4 represent the fixed fluid flow and respective profile of soil erosion without applying the algorithm of restructuring of the grid in the computational region.

Fig. 3. Flow around a pile without restructuring of a grid: velocity vector modulus.

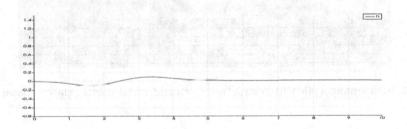

Fig. 4. Flow around a pile without restructuring of a grid: soil erosion boundaries.

Figures 5, 6 and 7 represent the development of fluid flow and respective profile of soil erosion with applying the algorithm of restructuring of the grid in the computational region.

The results of computation demonstrate the vital difference between algorithms without restructuring of the grid and in case of applying it. In the first case both fluid flow and soil erosion are rather quickly stabilized, as for another case we observe the formation of alluvion and its departing from the computational region with the course of time (ref. analogous behavior of erosion-accretion in [16]).

Fig. 5. Flow around a pile with restructuring of a grid: start of flow.

Fig. 6. Flow around a pile with restructuring of a grid: alluvion formation.

Fig. 7. Flow around a pile with restructuring of a grid: soil departing the computational region.

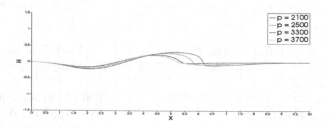

Fig. 8. Flow around a pile depending on density: start of flow.

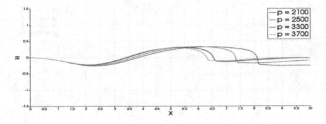

Fig. 9. Flow around a pile depending on density: alluvion formation.

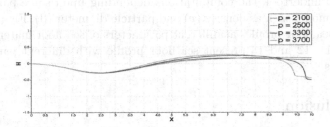

Fig. 10. Flow around a pile depending on density: departing the region.

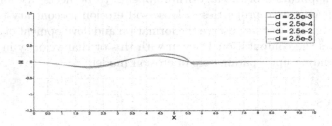

Fig. 11. Flow around a pile depending on particle size: start of flow.

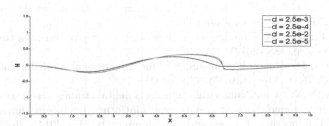

Fig. 12. Flow around a pile depending on particle size: alluvion formation.

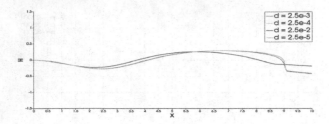

Fig. 13. Flow around a pile depending on particle size: departing the region.

Then we undertook the computations depending upon some parameters of soil erosion model such as density (p) and particle diameter (d). Figures 8, 9 and 10 present sea floor profiles at different parameters of sea floor material density.

Figures 11, 12 and 13 present sea floor profile with different parameters of soil particles.

5   Conclusion

The present paper contains numerical computations of the process of loose soil erosion applying the algorithm of restructuring of the finite-element grid for providing the feedback path with fluid flow. The computational results have shown that application of the algorithm appears to be necessary for obtaining the actual quantitative properties of loose soil erosion as contrary to the fixed grid because in this case we observe the formation and development of an alluvion which departs the computational region with this or that velocity in the course of time depending upon parameters of erosion model.

References

1. Chumakov, M.M., Onishchenko, D.A., Hahalina, S.N.: Soil erosion methodology near large ice formation keels. In: Gas Science News: Modern Approaches and Perspective Technologies in Projects of Oil-and-Gas Field Development of Russian shelf, Gazprom VNIIGAZ, Moscow, pp. 125–132 (2013)
2. Bonelli, S., Golay, F., Mercier, F.: Erosion of Geomaterials, pp. 187–222. Wiley/ISTE, Hoboken/Chicago (2012)
3. Mercier, F.: Numerical modelling of erosion of a cohesive soil by a turbulent flow. Ph.D. thesis, p. 188 (2013)
4. Lachouette, D., Golay, F., Bonelli, S.: One-dimensional modelling of piping flow erosion. C. R. de Mec. **336**, 731–736 (2008)
5. Zakharov, Y.N., et al.: Numerical and experimental studies of soil scour caused by currents near foundations of gravity-type platforms. In: International Conference on Civil Engineering, Energy and Environment (CEEE-2014), Hong Kong, pp. 190–197 (2014)
6. Zakharov, Y.N., et al.: Numerical and experimental studies of soil scour near foundations of platforms. In: Advanced Technologies of Hydroacoustics and Hydrophysics, pp. 239–241 (2014)

7. Zakharov, Y.N., et al.: Impact of waves and currents on the soil near gravity-type offshore platform foundation: numerical and experimental studies. In: Twenty-Fifth International Ocean and Polar Engineering Conference B "ISOPE-2015B", Kona, Hawaii, USA, pp. 807–814 (2015)
8. Zakharov, Y.N., Ivanov, K.S.: Computational investigation of turbulent flow impact on non-cohesive soil erosion near foundations of gravity type oil platforms. In: Mathematical and Information Tecnologies (MIT), Vrnjacka, Banja, Serbia, Budva, Montenegro, 18–5 August–September, pp. 524–534 (2016)
9. Zakharov, Y.N., Ivanov, K.S., Gaydarov, N.A.: Certificate of software state registration XFlow framework, 2015612750 (2015)
10. Belotserkovskii, O.M.: Numerical Simulation in Fluid Mechanics. Physmatlit, Moscow (1984)
11. Segal, I.A.: Finite Element Methods for the Incompressible Navier-stokes Equations. Delft University of Technology, Netherlands (2015)
12. Felippa, C.A., Guo, Q., Park, K.C.: Mass matrix templates: general description and 1D examples. Arch. Comput. Methods Eng. **22**(1), 1–65 (2015)
13. Saltykov, I.E.: Applying the finite element method for solving multi-dimensional unsteady Navier-Stokes equations. In: Fundamental and Applied Studies in Physics, Chemistry, Math and Informatics: Symposium Materials XII (XLIV) of International Theoretical and Practical Conference "Education, Science, Innovation: Young Researchers Contribution", pp. 324–327 (2017)
14. Zakharov, Y.N.: Gradient iterative methods for solving of hydrodynamics problems, Novosibirsk (2004)
15. Solberg, T., Hjertager, B.H., Bove, S.: CFD modelling of scour around offshore wind turbines in areas with strong currents. Technical Report 5, Esbjerg Institute of Technologym Aalborg University (2005)
16. Brørs, B.: Numerical modeling of flow and scour at pipelines. J. Hydraul. Eng. **125**, 511–523 (1999)

Stability Analysis of a Difference Scheme for the Dynamic Model of Gas Lift Process

B. Zhumagulov[1], N. Temirbekov[2], D. Baigereyev[3], and A. Turarov[3(✉)]

[1] National Engineering Academy of the Republic of Kazakhstan,
80 Bogenbay Batyr Street, 050010 Almaty, Kazakhstan
[2] Kazakhstan Engineering Technological University,
93 G/5 Al-Farabi, 050060 Almaty, Kazakhstan
[3] D. Serikbayev East Kazakhstan State Technical University,
19 Serikbayev Street, 070010 Ust-Kamenogorsk, Kazakhstan
t_ak17@mail.ru

Abstract. The paper studies a model of the gaslift process where the motion in a gas-lift well is described by partial differential equations. The system describing the studied process consists of equations of motion, continuity, equations of thermodynamic state, and concentration equation. A finite-difference scheme is constructed for the numerical solution of the problem. Because of the complexity of the motion equation, two simplified difference analogues for this equation are considered. The stability of these equations are investigated using the method of a priori estimates. The estimates obtained will be used for proving the stability of the whole difference scheme in future works.

Keywords: Gaslift process · Finite difference method ·
A priori estimates · Stability analysis · Motion equation · Simulation

1 Introduction

A rigorous justification of numerical methods for solving the problems of hydrodynamics is an urgent task in connection with its application in various fields of industry. One of such problems is the displacement of oil by the gaslift method, which is currently being used in many oil fields. This process is usually described by the motion equation

$$\varphi \rho_g \left(x, t \right) \left(\frac{\partial \overrightarrow{v}_g}{\partial t} + \overrightarrow{v}_g \cdot \nabla \overrightarrow{v}_g \right) + \nabla p(\rho) = -\frac{\lambda_c}{2d_g} \varphi \rho_g \overrightarrow{v}_g | \overrightarrow{v}_g | + \overrightarrow{f}_1, \quad (1)$$

$$\left(1 - \varphi \right) \rho_l \left(x, t \right) \left(\frac{\partial \overrightarrow{v}_l}{\partial t} + \overrightarrow{v}_l \cdot \nabla \overrightarrow{v}_l \right) + \nabla p(\rho) = -\frac{\lambda_c}{2d_g} \left(1 - \varphi \right) \rho_l \overrightarrow{v}_l | \overrightarrow{v}_l | + \overrightarrow{f}_2, \quad (2)$$

Y. Shokin and Z. Shaimardanov (Eds.): CITech 2018, CCIS 998, pp. 236–246, 2019.
https://doi.org/10.1007/978-3-030-12203-4_24

the equation of conservation of mass:

$$\frac{\partial \rho_\alpha}{\partial t} + \nabla \left(\rho_\alpha \vec{v}_\alpha \right) = 0, \quad \alpha = g, l, \tag{3}$$

the equation of concentrations:

$$\frac{\partial \varphi}{\partial t} + \vec{v}_l \cdot \nabla \varphi = f_3, \tag{4}$$

and the thermodynamic state equation

$$p = p(\rho), \tag{5}$$

$$\rho = \varphi \rho_g + (1 - \varphi) \rho_l, \tag{6}$$

in which subscripts g and l correspond to phases of gas and liquid, $\rho_\alpha(x, t)$ and $\vec{v}_\alpha(x, t)$ are the density and velocity of phase α, respectively, φ is concentration, p is pressure, $x \in \Omega \subset R^n$, $n = 1, 2, 3$. Let us denote $Q = [0, T] \times \Omega$. The coefficient of hydraulic resistance λ_c, the hydraulic channel diameter d_g are some positive constants, and the pressure $p(\rho)$ is a function of a positive argument with the first Lipschitz continuous derivative.

The system (1)–(5) is complemented with initial and boundary conditions:

$$\rho_\alpha|_{t=0} = \rho_{\alpha 0}(x), \quad \varphi|_{t=0} = \varphi_0(x), \quad \vec{v}_\alpha|_{t=0} = \vec{v}_{\alpha 0}(x), \tag{7}$$

$$\varphi|_S = \varphi_1(t), \quad \vec{v}_\alpha|_S = 0 \tag{8}$$

where $S = \partial \Omega$.

The system of equations describing the state of a viscous, compressible fluid, in contrast to the Eq. (1) also contains the viscous terms on the right-hand side. The solvability of the Cauchy problem on a small time interval for this system was studied in [1] and [2]. A theorem on the local solvability of the initial-boundary value problem for equations of a viscous, compressible fluid was proved in [3]. A local existence theorem for the solution of a one-dimensional initial-boundary value problem for the equations of motion of a viscous perfect polytropic gas in Lagrangian mass coordinates was proved in [4]. In [4] the technique of research of initial-boundary value problems "in the whole" on the time for the system of equations describing the one-dimensional flow of a viscous heat-conducting gas is described.

Various numerical methods for solving the systems of equations for the dynamics of a viscous compressible gas are currently used [5–11]. However, there is no mathematical proof of their stability and convergence to the solution of the differential problem. This is due to the nonlinearity of the equations, and also to the non-evolutionary nature of the system under consideration. For some problems of the dynamics of a viscous barotropic gas, an estimate of the error of difference schemes was obtained in the work of Kuznetsov and Smagulov [12,13]. A new difference scheme for the equations of a one-dimensional viscous heat-conducting gas is proposed and investigated in [14]. Studies of nonlinear difference schemes in the neighborhood of the known solution of specific problems of mathematical physics were carried out in [15,16].

In [17], the characteristics of various difference schemes for the Euler equations are compared for solutions of a number of model problems of gas dynamics and gas-dynamic processes. A new difference scheme for the nonstationary motion of a viscous barotropic gas in Euler variables is proposed in [18]. Positivity of the density function is ensured by the fact that not the values of the density function themselves are sought, but the natural logarithms of these quantities.

In the present paper, the Eqs. (1)–(7) are reduced to a equivalent form, and a difference scheme is constructed for solving the reduced problem. A priori estimates for velocities were obtained by the method of energy inequalities using methods applied in works [19–22].

2   Formulation of the Problem

Let us consider a simple one-dimensional problem in a segment $\Omega_1 \equiv [0,1]$ in which gas is injected through the left boundary. Using obvious transformations, the Eqs. (1)–(7) can be reduced to the following system of partial differential equations:

$$c(\varphi, \rho_g, \rho_l) \left(\frac{\partial v}{\partial t} + \frac{1}{2} \frac{\partial v^2}{\partial x} \right) + a(\varphi, \rho_l) v + \frac{\partial p}{\partial x} = -b(\varphi, \rho_g, \rho_l) v \, |v| , \qquad (9)$$

$$v_l = f(\varphi) v, \qquad (10)$$

$$\frac{\partial (\varphi \rho_g)}{\partial t} + \frac{\partial (\varphi \rho_g v)}{\partial x} = 0, \qquad (11)$$

$$\frac{\partial \varphi}{\partial t} + v_l \frac{\partial \varphi}{\partial x} = 0, \qquad (12)$$

$$p = \varphi p_g + (1 - \varphi) p_l, \qquad (13)$$

where $v \equiv v_g$,

$$c(\varphi, \rho_g, \rho_l) = \varphi \rho_g + (1 - \varphi) \rho_l f(\varphi),$$

$$a(\varphi, \rho_l) = (1 - \varphi) \rho_l \frac{\partial f(\varphi)}{\partial t},$$

$$f(\varphi) = \frac{0.83 - \varphi}{1 - \varphi},$$

$$b(\varphi, \rho_g, \rho_l) = \frac{\lambda_c}{2d_g} \left(\varphi \rho_g + (1 - \varphi) \rho_l f^2(\varphi) \right),$$

$x \in \Omega_1, t > 0$. Let us denote $Q_1 = [0, T] \times \Omega_1$.

Let us make some assumptions about coefficients in Eqs. (9)–(13).

1. Assume that oil is incompressible, i. e.

$$\rho_l = const, \quad \rho_l > 0. \tag{14}$$

2. Concentration φ is bounded:

$$0 \leq \varphi \leq 1. \tag{15}$$

3. f in (10) is non-negative for all values of concentration:

$$f(\varphi) \geq 0. \tag{16}$$

4. The gas density is a positive function:

$$\rho_g(p) > 0. \tag{17}$$

It follows from assumptions that $c(\varphi, \rho_g, \rho_l) > 0$.

Let us introduce a uniform difference grid $Q_h = \omega_\tau \times \omega_h$ with steps τ and h in Q_1. Let us assign the following difference scheme to the problem (9)–(13):

$$c_i^n v_{t,i}^n + \frac{1}{2} c_i^n v_{x,i}^n v_{x,i}^{n+1} + a_i^n v_i^n + p_{x,i}^n = -b_i^n v_i^{n+1} |v_i^n|, \tag{18}$$

$$v_{l,i} = f(\varphi_i) v_i, \tag{19}$$

$$(\varphi \rho_g)_{t,i} + (\varphi \rho_g v)_{\bar{x},i} = 0, \tag{20}$$

$$\varphi_{t,i} + v_{l,i} \varphi_{\bar{x},i} = 0, \tag{21}$$

$$p_i = \varphi_i p_{g,i} + (1 - \varphi_i) p_{l,i}. \tag{22}$$

Let us obtain a few a priori estimates for the difference motion Eq. (18).

Lemma 1. Under conditions (14)–(17) and

$$d_0 \leq c_i^n \leq d_1, \quad a_i^n \leq d_2, \quad b_i^n \leq d_3, \quad d_i > 0, \tag{23}$$

$$\frac{\varepsilon_3 d_1}{2} ||v_x^n||_C^2 + 2d_3 ||v^n||_C < \frac{d_0}{\tau} - \left(\frac{2d_1}{\varepsilon_3 h^2} + \frac{d_2}{\varepsilon_1} + \frac{1}{\varepsilon_2} \right) \tag{24}$$

the following estimate holds:

$$\nu \cdot ||v^{n+1}||^2 \leq M \cdot ||v^n||^2 + \tau \varepsilon_2 ||p_x^n||^2$$

where

$$\nu = d_0 - \tau \left(\frac{\varepsilon_3 d_1}{2} ||v_x^n||_C^2 + 2d_3 ||v^n||_C \right) - \tau \left(\frac{2d_1}{\varepsilon_3 h^2} + \frac{d_2}{\varepsilon_1} + \frac{1}{\varepsilon_2} \right).$$

Proof. Multiply (18) by $2\tau v_i^{n+1}$ and sum over nodes of the grid:

$$\left(c^n v_t^n,\ 2\tau v^{n+1}\right) + \left(\frac{1}{2}c^n v_x^n v_x^{n+1},\ 2\tau v^{n+1}\right) + \left(a^n v^n,\ 2\tau v^{n+1}\right)$$
$$+ \left(p_x^n,\ 2\tau v^{n+1}\right) = -\left(b^n v^{n+1}\,|v^n|,\ 2\tau v^{n+1}\right). \tag{25}$$

Estimate scalar products in (25) as follows:

$$\left|\left(c^n v_t^n,\ 2\tau v^{n+1}\right)\right| \geq d_0\left|\left(v_t^n,\ 2\tau v^{n+1}\right)\right| \geq d_0\left(||v^{n+1}||^2 - ||v^n||^2 + \tau^2||v_t^n||^2\right),$$

$$\left(\frac{1}{2}c^n v_x^n v_x^{n+1},\ 2\tau v^{n+1}\right) \leq \tau d_1 \max_{\omega_h}|v_x^n|\left|\left(v_x^{n+1},\ v^{n+1}\right)\right|$$

$$\leq \tau d_1 ||v_x^n||_C ||v_x^{n+1}|| \cdot ||v^{n+1}|| \leq \tau d_1\left(\frac{\varepsilon_3}{2}||v_x^n||_C^2 \cdot ||v^{n+1}||^2 + \frac{1}{2\varepsilon_3}||v_x^{n+1}||^2\right)$$

$$\leq \frac{\tau \varepsilon_3 d_1}{2}||v_x^n||_C^2 \cdot ||v^{n+1}||^2 + \frac{\tau d_1}{2\varepsilon_3} \cdot \frac{4}{h^2}||v^{n+1}||^2,$$

$$\left(a^n v^n,\ 2\tau v^{n+1}\right) \leq d_2\left|\left(v^n,\ 2\tau v^{n+1}\right)\right| = 2\tau d_2|(v^n,\ v^{n+1})| \leq 2\tau d_2||v^n|| \cdot ||v^{n+1}||$$

$$\leq 2\tau d_2\left(\frac{\varepsilon_1}{2}||v^n||^2 + \frac{1}{2\varepsilon_1}||v^{n+1}||^2\right),$$

$$\left(p_x^n,\ 2\tau v^{n+1}\right) \leq 2\tau\left|\left(p_x^n,\ v^{n+1}\right)\right| \leq 2\tau||p_x^n||\,||v^{n+1}||$$

$$\leq 2\tau\left(\frac{\varepsilon_2}{2}||p_x^n||^2 + \frac{1}{2\varepsilon_2}||v^{n+1}||^2\right),$$

$$-\left(b^n v^{n+1}\,|v^n|,\ 2\tau v^{n+1}\right) \leq 2\tau d_3|\left(v^{n+1}\,|v^n|,\ v^{n+1}\right)|$$

$$\leq 2\tau d_3 \max_{\omega_h}|v_x^n|\,|\left(v^{n+1},\ v^{n+1}\right)| = 2\tau d_3||v_x^n||_C \cdot ||v^{n+1}||^2.$$

It follows from (25) that

$$d_0\left(||v^{n+1}||^2 - ||v^n||^2 + \tau^2||v_t^n||^2\right) \leq \frac{\tau \varepsilon_3 d_1}{2}||v_x^n||_C^2 \cdot ||v^{n+1}||^2$$

$$+ \frac{\tau d_1}{\varepsilon_3} \cdot \frac{2}{h^2}||v^{n+1}||^2 + \tau d_2 \varepsilon_1||v^n||^2 + \frac{\tau d_2}{\varepsilon_1}||v^{n+1}||^2$$

$$+ \tau \varepsilon_2||p_x^n||^2 + \frac{\tau}{\varepsilon_2}||v^{n+1}||^2 + 2\tau d_3||v^n||_C \cdot ||v^{n+1}||^2,$$

$$\left(d_0 - \frac{\tau \varepsilon_3 d_1}{2}||v_x^n||_c^2 - \frac{2\tau d_1}{\varepsilon_3 h^2} - \frac{\tau d_2}{\varepsilon_1} - \frac{\tau}{\varepsilon_2} - 2\tau d_3||v^n||_c\right)||v^{n+1}||^2$$

$$+ d_0\tau^2||v_t^n||^2 \leq (d_0 + \tau d_2 \varepsilon_1)||v^n||^2 + \tau \varepsilon_2||p_x^n||^2,$$

Then

$$\nu \cdot ||v^{n+1}||^2 + d_0\tau^2||v_t^n||^2 \leq (d_0 + \tau d_2 \varepsilon_1) \cdot ||v^n||^2 + \tau \varepsilon_2||p_x^n||^2.$$

Thus, the assertion of the lemma holds under condition (24).

Let us neglect the concentration, and write the Eq. (1) in a non-divergent form taking into account the continuity Eq. (3) and the density positivity condition:

$$\frac{\partial v}{\partial t} + \frac{1}{2}\frac{\partial v^2}{\partial x} + \frac{\partial g}{\partial x} = -\frac{\lambda_c}{2d}v \cdot |v| + f \tag{26}$$

where $g = \ln \rho$. Then the Lax-Vendroff scheme for the Eq. (26) has the form

$$\frac{v_{i+\frac{1}{2}}^{n+\frac{1}{2}} - 0.5\left(v_{i+1}^n + v_i^n\right)}{\frac{\tau}{2}} + \frac{1}{2}\left[v_{i+1}^n \frac{v_{i+1}^n - v_i^n}{h} + v_i^n \frac{v_i^n - v_{i-1}^n}{h}\right]$$
$$+ \frac{g_{i+1}^n - g_i^n}{h} = -\frac{\lambda_c}{2d}|v_{i+\frac{1}{2}}^n| \cdot |v_{i+\frac{1}{2}}^n| + f_{i+\frac{1}{2}}^n \tag{27}$$

$$\frac{v_i^{n+1} - v_i^n}{\tau} + \frac{1}{2}\left[v_{i+\frac{1}{2}}^{n+\frac{1}{2}} \frac{v_{i+\frac{1}{2}}^{n+\frac{1}{2}} - v_{i-\frac{1}{2}}^{n+\frac{1}{2}}}{h} + v_{i-\frac{1}{2}}^{n+\frac{1}{2}} \frac{v_{i-\frac{1}{2}}^{n+\frac{1}{2}} - v_{i-\frac{3}{2}}^{n+\frac{1}{2}}}{h}\right]$$
$$+ \frac{g_{i+\frac{1}{2}}^{n+\frac{1}{2}} - g_{i-\frac{1}{2}}^{n+\frac{1}{2}}}{h} = -\frac{\lambda_c}{2d}|v_i^{n+\frac{1}{2}}| \cdot |v_i^{n+\frac{1}{2}}| + f_i^{n+\frac{1}{2}}. \tag{28}$$

Lemma 2. Under assumptions (14)–(17) the following estimate holds:

$$||v^{n+1}||^2 \le c||v^0||^2 + c\tau\left(||g||^2 + \sum_{k=0}^n \left(\left\|f^k\right\|^2 + \left\|f^{k+\frac{1}{2}}\right\|^2\right)\right).$$

Proof. Multiply the Eq. (27) by $\tau v_{i+\frac{1}{2}}^{n+\frac{1}{2}} h$ and sum along the inner nodes of the grid:

$$2\tau \sum_{i=1}^{M-1} \left(v_{i+\frac{1}{2}}^{n+\frac{1}{2}}\right)^2 h - 2\tau \sum_{i=1}^{M-1} v_{i+\frac{1}{2}}^{n} v_{i+\frac{1}{2}}^{n+\frac{1}{2}} h + \frac{\tau}{2} \sum_{i=1}^{M-1} v_{i+1}^n v_{x,i}^n v_{i+\frac{1}{2}}^{n+\frac{1}{2}} h$$
$$+ \frac{\tau}{2} \sum_{i=1}^{M-1} v_i^n v_{x,i}^n v_{i+\frac{1}{2}}^{n+\frac{1}{2}} h + \tau \sum_{i=1}^{M-1} g_{x,i}^n v_{i+\frac{1}{2}}^{n+\frac{1}{2}} h = -\frac{\lambda_c}{2d} \sum_{i=1}^{M-1} |v_{i+\frac{1}{2}}^n| v_{i+\frac{1}{2}}^{n+\frac{1}{2}} h \tag{29}$$
$$+ \tau \sum_{i=1}^{M-1} f_i^{n+\frac{1}{2}} v_{i+\frac{1}{2}}^{n+\frac{1}{2}} h.$$

Estimate scalar products in (29). Let us use the Cauchy inequality and the inequalities from [17] for the term generated by the nonlinear terms:

$$|j_1| \equiv \frac{\tau}{2}\left|\sum_{i=1}^{M-1} \left(v_{i+1}^n v_{x,i}^n + v_i^n v_{x,i}^n\right)\left(v_{i+\frac{1}{2}}^{i+\frac{1}{2}} - v_{n+\frac{1}{2}}^n\right)\right|$$
$$\le \frac{\tau}{h}|||v^n|^2|| \, ||v^{n+\frac{1}{2}} - v_n|| \le \frac{2\tau}{h}||v^n||^{\frac{1}{2}}||v_x^n||^{\frac{1}{2}}||v^{n+\frac{1}{2}} - v^n||$$
$$\le \frac{\varepsilon_1 \tau}{h}||v^{n+\frac{1}{2}} - v^n||^2 + \frac{2\tau}{\varepsilon_1 h^2}||v^n||^2,$$

similarly,

$$|j_2| \equiv \frac{\tau}{2}\left|\sum_{i=1}^{M-1} \left(v_{i+1}^n v_{x,i}^n + v_i^n v_{x,i}^n\right) v_{n+\frac{1}{2}}^n\right|$$
$$\le \frac{2\tau}{h}||v^n||^{\frac{1}{2}}||v_x^n||^{\frac{1}{2}}||v^n|| \le \frac{\tau}{h}\left(1 + \frac{2}{h}\right)||v^n||^2.$$

Using the difference analogue of the embedding theorem and ε-inequality to the term $j_3 \equiv \frac{\tau\lambda_c}{2d} \sum_{i=1}^{M-1} v_{i+\frac{1}{2}}^n |v_{i+\frac{1}{2}}^n| v_{i+\frac{1}{2}}^{n+\frac{1}{2}} h$, we obtain the inequality:

$$|j_3| \leq \frac{\tau\lambda_c}{2d}||v^n||^2||v^{n+\frac{1}{2}}|| \leq \frac{\varepsilon_1\tau\lambda_c}{4d}||v^n||^2 + \frac{1}{16\varepsilon_2}||v^n||^2||v_{\bar{x}}^{n+\frac{1}{2}}]|^2.$$

Using the formula of summation by parts to the last term on the left-hand side of (29), we obtain

$$|j_4| \leq \tau||v_x^{n+\frac{1}{2}}|| \, ||g|| \leq \frac{\varepsilon_1\tau}{2}||v_x^{n+\frac{1}{2}}||^2 + \frac{\tau}{2\varepsilon_1}||g||^2.$$

The term on the right-hand side of the Eq. (29) is estimated as follows:

$$|j_5| \leq \frac{\tau}{2}||v_x^{n+\frac{1}{2}}|| \, ||f^n|| \leq \frac{\varepsilon_1\tau}{16}||v_x^{n+\frac{1}{2}}||^2 + \frac{\tau}{8\varepsilon_1}||f^n||^2.$$

Substituting these inequalities into (29) and assuming that the inequalities $(\tau h - 2\varepsilon_1)/2h > 0$, $\frac{4}{h^2} - \frac{9\varepsilon_1\tau}{16} - \frac{1}{16\varepsilon_1}||v^n||^2 > 0$ hold, we obtain the inequality

$$||v_x^{n+\frac{1}{2}}||^2 \leq c_1||v^n||^2 + \frac{\tau}{2\varepsilon_1}||g||^2 + \frac{\tau}{8\varepsilon_1}||f^n||^2. \tag{30}$$

Similarly, multiply the Eq. (28) by $\tau v_i^{n+1} h$ and sum over the inner nodes of the grid:

$$||v^{n+1}||^2 + \tau^2||v_t^n||^2 + \frac{4}{h^2}||v_{\bar{x}}^{n+1}]|^2 + \frac{\tau}{2}\sum_{i=1}^{M-1}\left(v_{i+\frac{1}{2}}^{n+\frac{1}{2}} v_{x,i+\frac{1}{2}}^{n+\frac{1}{2}} + v_{i-\frac{1}{2}}^{n+\frac{1}{2}} v_{\bar{x},i-\frac{1}{2}}^{n+\frac{1}{2}}\right)v_i^n h$$

$$+\frac{\tau}{2}\sum_{i=1}^{M-1}\left(v_{i+\frac{1}{2}}^{n+\frac{1}{2}} v_{x,i+\frac{1}{2}}^{n+\frac{1}{2}} + v_{i-\frac{1}{2}}^{n+\frac{1}{2}} v_{\bar{x},i-\frac{1}{2}}^{n+\frac{1}{2}}\right)v_{t,i}^n h + \frac{\tau\lambda_c}{2d}|\sum_{i=1}^{M-1} v_i^{n+\frac{1}{2}}|v_i^{n+\frac{1}{2}}|v_i^{n+1} h$$

$$+\tau\sum_{i=1}^{M-1} v_i^{n+1} g_{x,i-\frac{1}{2}}^{n+\frac{1}{2}} h = 2||v^n||^2 + \tau\sum_{i=1}^{M-1} v_i^n f_i^{n+\frac{1}{2}} h. \tag{31}$$

Estimating the scalar products similarly to the Eq. (29), we obtain a similar estimate

$$||v^{n+1}||^2 \leq c_2||v^n||^2 + c_3\tau||g||^2 + c_3\tau||f^{n+\frac{1}{2}}||^2. \tag{32}$$

Adding the inequalities (30) and (32), we obtain

$$||v^{n+1}||^2 \leq c_4||v^n||^2 + c_5\tau||g||^2 + c_6\tau\left(||f^n||^2 + ||f^{n+\frac{1}{2}}||^2\right).$$

Using the inequality obtained, we can show that the chain of inequalities holds

$$||v^{n+1}||^2 \leq c_7||v^0||^2 + c_8\tau \, ||g||^2 + c_9\tau \sum_{k=0}^{n} \left(||f^k||^2 + \left\|f^{k+\frac{1}{2}}\right\|^2 \right),$$

which yields the assertion of the lemma.

3   The Results of Numerical Modeling

Using the algorithm above, a program was written to calculate the basic technological characteristics of the gas lift wells. The following input parameters were
set (Table 1):

Table 1. Initial data of the real deposit for modeling used in the numerical experiments

The value of	Units of measurement	Value
t - time	s	3600
L - length of wells	m	3496
ρ^g - is gas density	m/kg^3	0.75
ρ^l - is liquid density	m/kg^3	950
d_1 - is hydraulic diameter of the ring	m	0.0889
d_2 - is hydraulic diameter of the tubing	m	0.0759
D - is hydraulic diameter of the production string	m	0.168
p_p - injection pressure	MPa	9
p_r - reservoir pressure	MPa	19
p_w - wellhead pressure	MPa	1.5
T - temperature	K	333
g - is acceleration of gravity	m/s^2	9.80665

Figures 1, 2, 3 and 4 show the true content of the gas, density, pressure and
speed. Figure 4 shows the distribution of the true content of gas along the well.
Figure 5 shows the density distribution of gas, liquid, GLM along well. Figure 6
shows a function of pressure. It is seen from this graph that the fluid pressure in
the reservoir increases along the flow until the formation of gas-liquid mixture,
and then decreases. The graph of speed (Fig. 7) shows a monotonic decrease in
the rate of the medium to form a GLM. After the mixture of gas and liquid,
velocity in the tubing increases.

Figures 5, 6 and 7 on the left show the pressure distribution in a ring, and
figures on the right show the pressure distribution in tubing along the hole
depth at sections $n_1 = 100\tau$, $n_2 = 600\tau$, $n_3 = 1100\tau$, $n_4 = 2100\tau$, where the
dimensionless step time is set to $\tau = 10^{-3}$.

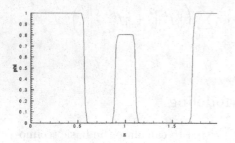

Fig. 1. The true content of the gas

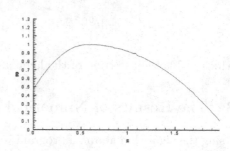

Fig. 2. Distribution of density

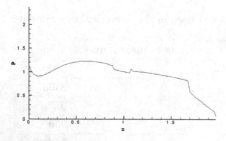

Fig. 3. Distribution of pressure

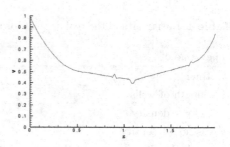

Fig. 4. Distribution of velocity

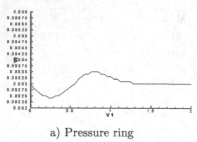

a) Pressure ring

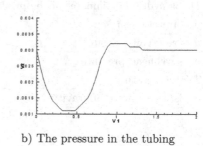

b) The pressure in the tubing

Fig. 5. The pressure distribution at $t = 100\tau$

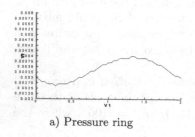

a) Pressure ring

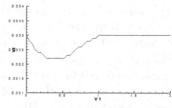

b) The pressure in the tubing

Fig. 6. The pressure distribution at $t = 600\tau$

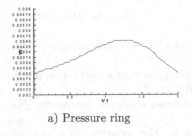

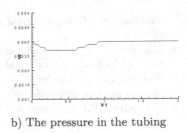

a) Pressure ring

b) The pressure in the tubing

Fig. 7. The pressure distribution at $t = 1100\tau$

4   Conclusion

Thus, in this paper the problem of fluid motion in a gas-lift well is considered. A difference scheme is proposed for the motion equation. Stability analysis is conducted for the phase velocities using the method of a priori estimates.

Using the proposed algorithm, a computer program is created for the numerical solution of one-dimensional problem for gas lift wells. The features of the numerical implementation of the algorithm for solving the one-dimensional problem are studied. A number of methodical calculations are conducted using the proposed numerical algorithms for solving the problem of determining the density, pressure, velocity in gas lift wells. It can be concluded from the results of calculations that the developed mathematical model, the difference scheme and the computer program allow to study the physical process in gas-lift wells and will be able to solve problems of optimal exploitation of oil deposits in future.

The obtained results can be used in studying the stability of difference schemes for solving the problem of displacement of oil by a gaslift method.

References

1. Nash, J.: Le probleme de Cauchy pour les equation differentielles d'un fluid general. Bull. Soc. Math. France **90**, 487–497 (1962)
2. Volpert, A.I., Khudyaev, S.I.: On the Cauchy problem for composite systems of nonlinear differential equations. Mat. Sb. **87**, 504–528 (1972)
3. Solonnikov, V.A.: On the solvability of the initial-boundary value problem for the equation of motion of a viscous compressible fluid. Invest. Linear Oper. Funct. Theory **6**, 128–142 (1976)
4. Antontsev, S.N., Kazhikhov, A.V., Monakhov, V.N.: Boundary value problems for the mechanics of inhomogeneous liquids. Science, Novosibirsk (1983)
5. Rozhdestvensky, B.L., Yanenko, N.N.: Systems of Quasilinear Equations and Their Applications to Gas Dynamics. Nauka, Moscow (1978)
6. Samarskii, A.A., Popov, Yu.P.: Difference methods for solving the problem of gas dynamics. Nauka, Moscow (1980)
7. Kovenya, V.M., Yanenko, N.N.: The splitting method in problems of gas dynamics. Science, Novosibirsk (1981)
8. Belotserkovsky, O.M., Davydov, Yu.M.: The method of large particles in gas dynamics. Nauka, Moscow (1982)

9. Fletcher, K.: Computational methods in the dynamics of liquids. Mir, Moscow (1991)
10. Godunov, S.K.: Difference method of numerical calculation of discontinuous solutions of hydrodynamics equations. Mat. Sb. **47**, 271–306 (1959)
11. Shokin, Yu.I., Yanenko, N.N.: The method of differential approximation: application to gas dynamics. Science, Novosibirsk (1985)
12. Kuznetsov, B.G., Smagulov, Sh.: On convergent difference schemes for viscous gas equations, Novosibirsk (1982)
13. Smagulov, S.: On convergent difference schemes for the equations of a viscous heat-conducting gas. Dokl. Akad. Nauk SSSR **275**, 31–34 (1984)
14. Amosov, A.A., Zlotnik, A.A.: A difference scheme for the equations of motion of a viscous heat-conducting gas, its properties and error estimates "in the large". DAN SSR **284**, 265–269 (1985)
15. Abashin, V.N.: Difference schemes of gas dynamics. Differ. Equ. **17**, 710–718 (1981)
16. Lyashko, A.D., Fedorov, E.M.: On the correctness of nonlinear two-layered operator-difference schemes. Differ. Equ. **17**, 1304–1316 (1981)
17. Volkov, K.N.: Difference schemes for calculating flows of high resolution and their application for solving gas dynamics problems. Comput. Methods Program. **6**, 146–167 (2005)
18. Popov, A.V., Zhukov, K.A.: An implicit difference scheme for the nonstationary motion of a viscous barotropic gas. Comput. Methods Program. **14**, 516–623 (2013)
19. Temirbekov, N.M., Turarov, A.K., Baigereyev, D.R.: Numerical modeling of the gas lift process in gas lift wells. In: AIP Conference Proceedings, vol. 1739, Article id 020067 (2016)
20. Temirbekov, A.N., Wójcik, W.: Numerical implementation of the fictitious domain method for elliptic equations. Int. J. Electron. Telecommun. **60**(3), 219–223 (2014)
21. Temirbekov, A.N.: Numerical implementation of the method of fictitious domains for elliptic equations. In: AIP Conference Proceedings, vol. 1759, Article id 020053 (2016)
22. Zhumagulov, B., Temirbekov, N., Baigereyev, D.: Efficient difference schemes for the three-phase non-isothermal flow problem. In: AIP Conference Proceedings 1880, Article id 060001 (2017)

Author Index

Printed in the United States
By Bookmasters